T0202904

Communications in Computer and Information Science 1940

Rationale

The CCIS series is devoted to the publication of proceedings of computer science conferences. Its aim is to efficiently disseminate original research results in informatics in printed and electronic form. While the focus is on publication of peer-reviewed full papers presenting mature work, inclusion of reviewed short papers reporting on work in progress is welcome, too. Besides globally relevant meetings with internationally representative program committees guaranteeing a strict peer-reviewing and paper selection process, conferences run by societies or of high regional or national relevance are also considered for publication.

Topics

The topical scope of CCIS spans the entire spectrum of informatics ranging from foundational topics in the theory of computing to information and communications science and technology and a broad variety of interdisciplinary application fields.

Information for Volume Editors and Authors

Publication in CCIS is free of charge. No royalties are paid, however, we offer registered conference participants temporary free access to the online version of the conference proceedings on SpringerLink (http://link.springer.com) by means of an http referrer from the conference website and/or a number of complimentary printed copies, as specified in the official acceptance email of the event.

CCIS proceedings can be published in time for distribution at conferences or as post-proceedings, and delivered in the form of printed books and/or electronically as USBs and/or e-content licenses for accessing proceedings at SpringerLink. Furthermore, CCIS proceedings are included in the CCIS electronic book series hosted in the SpringerLink digital library at http://link.springer.com/bookseries/7899. Conferences publishing in CCIS are allowed to use Online Conference Service (OCS) for managing the whole proceedings lifecycle (from submission and reviewing to preparing for publication) free of charge.

Publication process

The language of publication is exclusively English. Authors publishing in CCIS have to sign the Springer CCIS copyright transfer form, however, they are free to use their material published in CCIS for substantially changed, more elaborate subsequent publications elsewhere. For the preparation of the camera-ready papers/files, authors have to strictly adhere to the Springer CCIS Authors' Instructions and are strongly encouraged to use the CCIS LaTeX style files or templates.

Abstracting/Indexing

CCIS is abstracted/indexed in DBLP, Google Scholar, EI-Compendex, Mathematical Reviews, SCImago, Scopus. CCIS volumes are also submitted for the inclusion in ISI Proceedings.

How to start

To start the evaluation of your proposal for inclusion in the CCIS series, please send an e-mail to ccis@springer.com.

Akram Bennour · Ahmed Bouridane ·
Lotfi Chaari

Editors

Intelligent Systems and Pattern Recognition

Third International Conference, ISPR 2023
Hammamet, Tunisia, May 11–13, 2023
Revised Selected Papers, Part I

Springer

Editors
Akram Bennour 🆔
Larbi Tebessi University
Tebessa, Algeria

Ahmed Bouridane 🆔
Sharjah University
Sharjah, United Arab Emirates

Lotfi Chaari 🆔
University of Toulouse
Toulouse, France

ISSN 1865-0929 ISSN 1865-0937 (electronic)
Communications in Computer and Information Science
ISBN 978-3-031-46334-1 ISBN 978-3-031-46335-8 (eBook)
https://doi.org/10.1007/978-3-031-46335-8

This Springer imprint is published by the registered company Springer Nature Switzerland AG
The registered company address is: Gewerbestrasse 11, 6330 Cham, Switzerland

Paper in this product is recyclable.

Preface

We are delighted to present the proceedings of the ISPR 2023: The Third International Conference on Intelligent Systems and Pattern Recognition. This event was meticulously organized by the Artificial Intelligence and Knowledge Engineering Research Labs at Ain Sham University in collaboration with the MIRACL laboratory at Sfax University, Tunisia. The conference served as a dynamic platform for interdisciplinary discourse, facilitating the exchange of cutting-edge developments across various domains of artificial intelligence and pattern recognition. Supported by the esteemed International Association of Pattern Recognition (IAPR), the conference took place in the picturesque locale of Hammamet, Tunisia, from May 11–13, 2023.

Within this compilation of proceedings lies a collection of papers that have been meticulously vetted and showcased during the conference. The conference drew an impressive array of scholarly contributions, with a grand total of 129 papers submitted across diverse domains within pattern recognition and artificial intelligence. These submissions underwent rigorous evaluation, undertaken by esteemed researchers hailing from various corners of the globe, each an authority in their respective fields. Following a thorough and competitive at least 3 reviews per submission in a double-blind review, 49 outstanding papers emerged triumphant, earning the opportunity to grace the conference podium. Ultimately, 44 of these distinguished papers were registered and adeptly presented and discussed during the event, attaining a rate of 34%.

We extend our heartfelt gratitude to the diligent reviewers, whose invaluable time and dedication were instrumental in evaluating the papers and offering insightful feedback to the authors. Our appreciation also extends to the esteemed keynote speakers, authors, and participants who collectively contributed to the conference's success. We extend our recognition to authors whose submissions were not selected this time; they are encouraged to explore the potential inclusion of their research papers in the forthcoming edition of ISPR.

The relentless efforts of the organizing committees have undoubtedly played a pivotal role in orchestrating this remarkable event, and for that, they deserve resounding praise.

A special acknowledgment goes to the International Association of Pattern Recognition (IAPR) and the event's sponsors, whose support has significantly enriched the conference's stature. We trust that ISPR 2023 fostered invaluable knowledge exchange and networking avenues for all participants. We eagerly anticipate your presence once again at the next edition.

August 2023

Akram Bennour
Ahmed Bouridane
Lotfi Chaari

Organization

General Chairs

Akram Bennour	Larbi Tebessi University, Algeria
Tolga Ensari	Arkansas Tech University, USA
Abdel-Badeeh Salem	Ain Shams University, Egypt

Steering Committee

Akram Bennour	Larbi Tebessi University, Algeria
Lotfi Chaari	INP-Toulouse, France
Moises Diaz	Universidad de Las Palmas de Gran Canaria, Spain
Bechir Alaya	Gabes University, Tunisia
Najib Ben Aoun	Al Baha University, Saudi Arabia
Abdel-Badeeh Salem	Ain Shams University, Egypt
Yahia Slimani	Manouba University, Tunisia
Atta ur Rehman Khan	Ajman University, UAE

International Advisory Board

Mohamed Elhoseny	University of Sharjah, UAE
K. C. Santosh	University of South Dakota, USA
José Ruiz-Shulcloper	University of Informatics Sciences, Cuba
Linas Petkevičius	Vilnius University, Lithuania
Mohammed al-Sarem	Tayba University, Saudi Arabia
Faiz Gargouri	Sfax University, Tunisia
Takashi Matsuhisa	Karelia Research Centre, Russian Academy of Science, Russia
Sabah Mohammed	Lakehead University, Canada
Mostafa M. Aref	Ain Shams University, Egypt

Program Chairs

Akram Bennour	Larbi Tebessi University, Algeria
Ahmed Bouridane	Sharjah University, UAE
Lotfi Chaari	INP-Toulouse, France

Publicity Chairs

Ahmed Cheikh Rouhou	National Engineering School of Sfax, Tunisia
Chintan M. Bhatt	Pandit Deendayal Energy University, India
Mohammed al-Chaabi	Tayba University, Saudi Arabia
Sean Eom	Southeast Missouri State University, USA
Ali Ismail Awad	United Arab University, Abu Dhabi, UAE
Majid Banaeyan	TU Wien, Austria
Mustafa Dagtekin	Istanbul University - Cerrahpaşa, Turkey
Mohamed Hammad	Menoufia University, Egypt
Mustafa Ali Abuzaraida	Utara University, Malaysia
Jinan Fiaidhi	Lakehead University, Canada

Sponsors and Exhibitions Chairs

Samir Tag	Larbi Tebessi University, Algeria
Hatem Haddad	Institut Supérieur des Arts Multimédia de la Manouba, Tunisia
Fedoua Drira	National Engineering School of Sfax, Tunisia
Gattel Abdejalil	Larbi Tebessi University, Algeria

Local Arrangement Committee

Bassem Bouaziz (LAC Chair)	University of Sfax, Tunisia
Adham Bennour	Constantine University, Algeria
Takwa Benaicha	Tunis University, Tunisia
Marzoug Soltan	Larbi Tebessi University, Algeria
Saméh Kchaou	University of Sfax, Tunisia
Walid Mahdi	University of Sfax, Tunisia
Yousra Bendaly Hlaoui	University of Tunis El Mana, Tunisia
Walid Barhoumi	University of Carthage, Tunisia

Publication Chairs

Tolga Ensari	Arkansas Tech University, USA
Mohamed Elhoseny	University of Sharjah, UAE
Mohammed al-Sarem	Tayba University, Saudi Arabia
Mohamed Hammad	Menoufia University, Egypt
Sajid Anwar	Institute of Management Sciences, Pakistan
Najib Ben Aoun	Al Baha University, Saudi Arabia
Varuna De Silva	Loughborough University, UK
Imad Rida	University of Technology of Compiègne, France

Program Committee

Niketa Gandhi	MIR Labs, USA
Souad Guessoum	Badji Mokhtar University, Annaba, Algeria
Takashi Matsuhisa	Institute of Applied Mathematical Research, Karelia Research Centre, Russia
Ehlem Zigh	USTO, Algeria
Hatem Haddad	Institut Supérieur des Arts Multimédia de la Manouba, Tunisia
Mousa Albashrawi	King Fahd University of Petroleum and Minerals, KSA
Djeddar Afrah	Saad Dahleb University, Blida, Algeria
Yaâcoub Hannad	National Institute of Statistics and Applied Economics, Morocco
Varuna De Silva	Loughborough University, UK
Mohamed Elhoseny	University of Sharjah, UAE
Mohamed Ben Halima	Institut Supérieur d'Informatique et de Multimédia de Sfax, Tunisia
Karri Chiranjeevi	University of Porto, Portugal
Faiz Gargouri	University of Sfax, Tunisia
Linas Petkevičius	Vilnius University, Lithuania
Mohammed Benmohammed	Constantine University, Algeria
Faouzi Ghorbel	CRISTAL/ENSI, Tunisia
José Ruiz-Shulcloper	University of Informatics Sciences, Cuba
Mohammed al-Sarem	Tayba University, Saudi Arabia
Chintan M. Bhatt	Pandit Deendayal Energy University, India
K. C. Santosh	University of South Dakota, USA
Mohamed Hammad	Menoufia University, Egypt
Idris El-Feghi	SomaDetect, Canada
Bassem Bouaziz	University of Sfax, Tunisia

Najib Ben Aoun	Al Baha University, Saudi Arabia
Hemen Dutta	Gauhati University, India
Sirine Marrakchi	University of Sfax, Tunisia
Ali Abu Odeh	University of Technology, Bahrain
Salma Jamoussi	University of Sfax, Tunisia
Emna Fendri	Miracl/FSS, Tunisia
Ljubica Kazi	University of Novi Sad, Serbia
Ali Ismail Awad	United Arab Emirates University, UAE
Imran Siddiqi	Bahria University, Pakistan
Sabah Mohammed	Lakehead University, Canada
Sumaya Al Maadeed	Qatar University, Qatar
Azza Ouled Zaid	ISI, Tunisia
Jinan Fiaidhi	Lakehead University, Canada
Lotfi Chaari	INP-Toulouse, France
Sugata Gangopadhyay	Indian Institute of Technology Roorkee, India
Mohamed Hammami	Miracl/FSS, Tunisia
Ahmed Bouridane	Sharjah University, UAE
Imad Rida	University of Technology of Compiègne, France
Moises Diaz	Universidad de Las Palmas de Gran Canaria, Spain
Mustafa Ali Abuzaraida	Utara University, Malaysia
Yaâcoub Hannad	Mohammed V University of Rabat, Morocco
Mohammed al-Chaabi	Tayba University, Saudi Arabia
Xuewen Yang	InnoPeak Technology Inc., USA
Mustafa Dagtekin	Istanbul University - Cerrahpaşa, Turkey
Yahia Slimani	Manouba University, Tunisia
Hammad Afzal	National University of Sciences and Technology, Pakistan
Nemmour Hassiba	USTHB, Algeria
Vijayan K. Asari	University of Dayton, USA
Atta ur Rehman Khan	Ajman University, UAE
Slim M'Hiri	CRISTAL/ENSI, Tunisia
Sam Zaza	Middle Tennessee State University, USA
Bechir Alaya	Gabes University, Tunisia
Rachid Oumlil	École Nationale de Commerce et de Gestion, Morocco
Mahnane Lamia	Badji Mokhtar University, Annaba, Algeria
Tolga Ensari	Arkansas Tech University, USA
Fedoua Drira	National Engineering School of Sfax, Tunisia
Jerry Wood	Arkansas Tech University, USA
Toufik Sari	Badji Mokhtar University, Annaba, Algeria
Sean Eom	Southeast Missouri State University, USA

Walid Mahdi	University of Sfax, Tunisia
Anna-Maria Di Sciullo	UQAM, Canada
Bhaskar Ghosh	Arkansas Tech University, USA
Abdelkader Nasreddine Belkacem	Osaka University, Japan
Imed Riadh Ferah	RIADI/MSE, Tunisia
Faezeh Soleimani	Ball State University, USA
Ezeddine Zagrouba	ISI, Tunisia
Robin Ghosh	Arkansas Tech University, USA
Indira Dutta	Arkansas Tech University, USA
Suzan Anwar	Philander Smith College, USA
Nassima Bouchareb	Constantine University, Algeria
Shridhar Devamane	Global Academy of Technology, India
Hafidi Mohamed	Cadi Ayyad University, Morocco
Sid Ahmed Benabderrahmane	Inria, France
Nadia Zeghib	University of Constantine 2 - Abdelhamid Mehri, Algeria
Shahin Gelareh	University of Artois, France
Abdel-Badeeh Salem	Ain Shams University, Egypt
Nacereddine Zarour	Constantine 2 University, Algeria
Xinli Xiao	Arkansas Tech University, USA
Amel Benazza	Sup'Com, Tunisia
Weiru Chen	Arkansas Tech University, USA
Yudith Cardinale	Simón Bolívar University, Miranda, Venezuela
Mebarka Yahlali	Saida University, Algeria
Ankur Singh Bist	TowardsBlockchain, India
El-Sayed M. El-Horbaty	Ain Shams University, Egypt
Abdelghani Ghomari	University of Oran1, Algeria
Dragana Krstić	University of Niš, Serbia
Kechar Bouabdellah	University of Oran, Algeria
Krassimir Markov	Institute of Information Theories and Applications, Bulgaria
Dana Simian	Lucian Blaga University, Romania
Marina Nehrey	National University of Life and Environmental Sciences of Ukraine, Ukraine
Rossitsa Yalamova	University of Lethbridge, Canada
Ouassila Hioual	Constantine 2 University, Algeria
Vitalina Babenko	Kharkiv National University, Ukraine
El-Sayed A. El-Dahshan	Ain Shams University, Egypt
Mostafa M. Aref	Ain Shams University, Egypt
Vera Meister	Brandenburg University of Applied Sciences, Germany
Paata Kervalishvili	Georgian Technical University, Georgia

Walid Barhoumi	University of Carthage, Tunisia
Roumen Kountchev	Technical University of Sofia, Bulgaria
Nagwa Badr	Ain Shams University, Egypt
Rasha Ismail	Ain Shams University, Egypt
Natalya Shakhovska	Lviv Polytechnical National University, Ukraine
Anastasia Y. Nikitaeva	Southern Federal University, Russia
Francesco Sicurello	University of Milan-Bicocca, Italy
Maria Brojboiu	University of Craiova, Romania
Liliana Moga	Dunarea de Jos University of Galati, Romania
Nouhad Rizk	University of Houston, USA
Volodymyr Romanov	Glushkov Institute of Cybernetics of National Academy of Sciences of Ukraine, Ukraine
Romina Kountchev	Technical University of Sofia, Bulgaria
Cornelia Aida Bulucea	University of Craiova, Romania
Elena Nechita	Vasile Alecsandri University, Romania
Vicente Rodríguez Montequín	Universidad de Oviedo, Spain
Livia Bellina	MobileDiagnosis Onlus, Italy
Yukako Yagi	Harvard Medical School, USA
Qinghan Xiao	Defence R&D, Canada
Felix T. S. Chan	Macau University of Science and Technology, China
Redouane Tlemsani	University of Sciences and Technology of Oran, Algeria
Sami Saleh	Sains University, Malaysia
Djamel Samai	Ouargla University, Algeria
Sourour Ammar	Digital Research Center of Sfax, Tunisia
Yousra Bendaly Hlaoui	University of Tunis El Mana, Tunisia
Paulo Batista	University of Évora, Portugal
Baskar Arumugam	Amrita Vishwa Vidyapeetham University, India
Loukas Ilias	National Technical University of Athens, Greece
Ghalem Belalem	Université d'Oran, Algeria
Nabiha Azizi	Badji Mokhtar University, Annaba, Algeria
Mouna Rekik	Sousse University, Tunisia
Gattel Abdejalil	Larbi Tebessi University, Algeria
Leïla Boussaad	University of Batna, Algeria
Rebiha Zeghdane	University of Bordj Bou Arreridj, Algeria
Imran Mudassir	Air University Islamabad, Pakistan
Kais Khrouf	Jouf University, KSA
Nassima Aissani	University of Oran, Algeria
Abdellatif Rahmoun	ESI, Algeria
Djamila Mohdeb	University of Bordj Bou Arreridj, Algeria
Rudresh Dwivedi	Netaji Subhas University of Technology, India

Amel Hebboul   École Normale Supérieure de Constantine,
                                   Algeria
Adel Alti   University of Setif, Algeria
Djakhdjakha Lynda   Université 8 Mai 1945 Guelma, Algeria
Mariagrazia Fugini   Politecnico di Milano, Italy
Akram Boukhamla   Badji Mokhtar University, Algeria
Arcangelo Castiglione   University of Salerno, Italy
Chaouki Boufenar   Inria Saclay Ile-de-France, France
Mohamed Jmaiel   University of Sfax, Tunisia
Kawther Abbas-Sallal   University of Technology and Applied Sciences,
                                   Oman
Osvaldo Gervasi   University of Perugia, Italy
Andrea F. Abate   University of Salerno, Italy
Belbachir Khadidja   USTO MB University, Algeria
Javad Sadri   McGill University, Canada
Youcef Chibani   USHTB, Algeria
Zakaria Maamar   Zayed University, UAE
Hanene Trichili   University of Sfax, National School of Engineers
                                   (ENIS), Tunisia
Chutisant Kerdvibulvech   National Institute of Development
                                   Administration, Thailand
Rahma Boujelbane   University of Sfax, Tunisia
Kouidri Siham   Saida University, Algeria
Hazem Abbas   Queen's University, Canada
Daniel Arockiam   Galgotias University, India
Salem Nasri   Qassim University, KSA
Chikh Mohammed Amine   Tlemcen University, Algeria

Contents – Part I

Data Mining

Contents – Part II

Computer Vision

Impact of Neural Network Architecture for Fingerprint Recognition

Simon Hallösta$^{(\boxtimes)}$, Mats I. Pettersson, and Mattias Dahl

Blekinge Institute of Technology, Biblioteksgatan 8, 374 32 Karlshamn, Blekinge,
Sweden
{simon.hallosta,mats.pettersson,mattias.dahl}@bth.se

Abstract. This work investigates the impact of the neural networks architecture when performing fingerprint recognition. Three networks are studied; a Triplet network and two Siamese networks. They are evaluated on datasets with specified amounts of relative translation between fingerprints. The results show that the Siamese model based on contrastive loss performed best in all evaluated metrics. Moreover, the results indicate that the network with a categorical scheme performed inferior to the other models, especially in recognizing images with high confidence. The Equal Error Rate (EER) of the best model ranged between $4\% - 11\%$ which was on average 6.5 percentage points lower than the categorical schemed model. When increasing the translation between images, the networks were predominantly affected once the translation reached a fourth of the image. Our work concludes that architectures designed to cluster data have an advantage when designing an authentication system based on neural networks.

Keywords: Fingerprint recognition · Neural network architecture · Siamese network

1 Introduction

The fingerprint has been one of the most prominent biometric features used to identify individuals for many decades in forensics. With the introduction of fingerprint readers on mobile devices, the need for efficient and highly reliable matching algorithms has increased [6]. Since the fingerprint readers on mobile devices are relatively small, the problem of fingerprint matching of partial fingerprints arises. Even though the overlap between fingerprint readings is significant, the feature density in the captured region can make it challenging to make an accurate identification [16].

Fingerprint features can be grouped into three levels [14]. Level-1 features consist of overall structure such as swirl and center structure. Level-2 is on ridge level, such as minutiae. Level-3 features describe the sub-ridge pattern features, e.g., pores and scares. The more features an image contains, the better performance is to be expected. The level of features that is possible to extract depends on the sensor's resolution. With capacitive sensors, a resolution of 500

A. Bennour et al. (Eds.): ISPR 2023, CCIS 1940, pp. 3–14, 2024.
https://doi.org/10.1007/978-3-031-46335-8_1

ppi can be achieved. At this resolution level, 1 and 2 are feasible to extract. With resolutions greater than 800 ppi, level-3 features become recognizable [14].

In the last decade, there have been tremendous advances in the field of computer vision and pattern recognition due to the use of deep learning. This technique has shown excellent results in various problems such as classification, object detection, segmentation, face recognition, and many more. In fingerprint recognition the objective is to measure the similarity between images of fingerprints; this is very similar to the task in face recognition. As shown in [10,11], neural networks in Siamese network architectures have proven to be very capable of measuring the similarity between faces. Since then, many variants of Siamese networks have been used in fingerprint recognition [1,2,4,15,17,18].

When using neural networks, it can be hard to determine what information in the input led to the final decision taken by the network. Traditionally, the minutiae features in fingerprints have been one of the most widely used techniques to determine matching fingerprints. In [4] they showed that a Siamese neural network without domain knowledge of the concept of minutiae features still established the significance of minutiae.

With support of the location of minutiae, promising results were achieved using Triplet networks [11] for matching patches in conjunction with a global fingerprint matching scheme [15,17]. Their study was based on high-resolution fingerprints. A similar approach applied to high-resolution images was developed in [2], where they focused on the level-3 feature, pores. They also established matching patches around pores using a Triplet network.

In [18], they used Siamese networks on lower resolution fingerprints. They trained their networks to perform similarity measurements using a contrastive loss function [5]. Three models utilizing the same loss function were compared. Their work indicated that the choice of model for feature extraction had an eminent influence on the resulting accuracy of the networks.

There exists a multitude of different techniques and sensor types to obtain fingerprint data. In [1], they investigate how a Siamese network based on cross-entropy and an adversarial network could handle cross-sensor matching. The adversarial part of the network forced the model to learn to extract non-sensor-specific features. Some of the sensors used in their work bear great resemblance to the sensor used in our experiments, i.e., a capacitive sensor with a resolution of 500 ppi.

In our work, we implement the core ideas of three of the approaches described above, most closely related to [1,15,18]. We then proceed to investigate how the architecture of neural networks impacts the final performance and how they respond to relative translation between partial fingerprints.

2   Method

When developing a matching algorithm for fingerprints, a strong incentive is to be able to learn new fingerprints with a small amount of data. Making neural networks label each user as its own class is not desirable due to the need to retrain the network with each new fingerprint entered into the database. To circumvent this issue, architectures that can infer matching as a label or cluster the fingerprints in feature space are sought after. One of these architectures is the Siamese network which has shown great success in facial recognition tasks [10].

2.1   Contrastive Model

A Siamese network has two branches that process the input to the network, see Fig. 1. The two branches usually share some or all of their weights, and the output is then compared. The comparison can be made by incorporating it into a loss function, i.e., Contrastive loss [5],

$$L_C = \frac{1}{N} \sum_{k=1}^{N} \frac{1}{2} l_k \left\| \mathbf{B}(\mathbf{i}_k^1) - \mathbf{B}(\mathbf{i}_k^2) \right\|_2^2 + \frac{1}{2}(1 - l_k) \left[\max(0, \ m - \left\| \mathbf{B}(\mathbf{i}_k^1) - \mathbf{B}(\mathbf{i}_k^2) \right\|_2^2) \right], \quad (1)$$

where $\mathbf{i}_k^{1,2}$ is the k-th image pair in a batch, N is the number of image pairs in the batch, l_k is 1 if the pair matches otherwise, 0. The function $\mathbf{B}$ is the normalized output from the network. The hyperparameter m is the margin where non-matching fingerprints contribute to the loss. The model based on the Contrastive loss is denoted **Contrastive** model from now on.

2.2   Decision Model

Another way of training a Siamese network is to combine the output of branches inside the network and let the network learn how to classify feature embeddings. In our case, the two branches are combined using element-wise subtraction of a vector representation of the features generated by the base model, see Fig. 1. The vector is then passed to fully connected layers, which functions as a classifier. The network is then trained using a cross-entropy loss function with two classes, matching and non-matching [1],

$$L_D = \frac{1}{N} \sum_{k=1}^{N} l_k \log(p_k) + (1 - l_k) \log(1 - p_k), \quad (2)$$

with p_k denoting the probability of image pair k to be matching. This model will be referred to as the **Decision** model.

2.3   Triplet Model

The Triplet architecture has inherited a lot of the Siamese networks' structure but adds a third branch. As with the Siamese networks, the weights of the base models are shared between these branches, see Fig. 1. The distinction is in the loss function and input during training. Then the input is divided into three; anchor, positive and negative examples. Where the positive fingerprint is matching to the anchor and the negative is non-matching. The network is trained using the Triplet loss function [11]

$$L_T = \frac{1}{N} \sum_{k=1}^{N} \max \left(0, \ \|\mathbf{B}(\mathbf{i}_k^a) - \mathbf{B}(\mathbf{i}_k^p)\|_2^2 - \|\mathbf{B}(\mathbf{i}_k^a) - \mathbf{B}(\mathbf{i}_k^n)\|_2^2 + \alpha \right), \qquad (3)$$

where $\mathbf{i}_k$ is the k:th image triplet in a batch, the superscripts p, a, n denote positive, anchor and negative images, respectively. The function $\mathbf{B}$ is the normalized output from the network, and N is the total number of triplets in a training batch. The α parameter is a hyperparameter functioning as a margin to only contribute hard triplets to the loss of each batch. The model utilizing this loss function will henceforth be identified as the **Triplet** model.

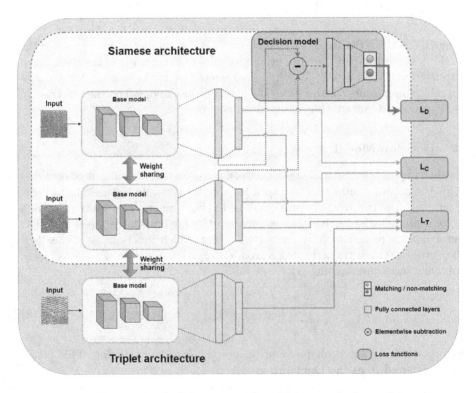

Fig. 1. Illustration of all three network architectures during training.

Due to the security nature of biometric authentication, the systems need to be both reliable and secure at a very high confidence level. This entails that the most important information is learned from the most challenging pairs of fingerprints. Hence, all models utilize online hard example mining [11] to back-propagate the loss from the hardest image pairs. This is done by top-k filtering of the computed loss for each image pair in a batch. In the case of the decision layer model, this is done by first calculating the L_D loss, see Eq. 2, individually for each pair and then propagating the mean of the k largest loss contributions. When performing hard example mining in the Contrastive model, an equal split of the k hardest matching and k hardest non-matching pairs contributes to the final loss value. There is an additional hyperparameter controlling how many extra negative examples should be considered for the Triplet model for every positive pair.

At inference time, the feature embeddings of the users' template images have been stored. This enables the possibility to only use one branch in the Siamese and triplet networks when authenticating users. Thus, only the distance measured between the feature embedding of the incoming image and the stored template must be computed. This makes for an inherent safe storage alternative for users' biometric data. Finally, a threshold level set from validation data determines the decision to authorize a user into the system.

All models relied upon a transfer learning from pre-trained networks, referred to as base models see Fig. 1, to extract features from the images. The base models used during the hyperparameter search were EfficientNet [13], and ResNet [7] models. After feature extraction, the models used fully connected layers before either using the final feature embedding in the loss function or joining the two branches, depending on the model investigated.

With the aim of finding the best-performing networks, Bayesian optimization was used to search the hyperparameter space [12]. Hyperparameters included: pre-trained base model, learning rate, parameters in fully connected layers, number of trainable layers, parameters for online hard example mining, and loss function parameters. In order to reduce the time spent training, early stopping was applied for unpromising models.

3   Results

3.1   Dataset

The performance assessment and model training was conducted utilizing a dataset that is proprietary in nature, as referenced in [8]. The fingerprint dataset is composed of partial fingerprint images with a resolution of $[192 \times 192]$ pixels. The images are from a $[10 \times 10]$ millimeter capacitive sensor with 508 pixels per inch (ppi). Each image is accompanied by alignment information in the form of its rotation and translation relative to other images of the same fingerprint. As convolutional neural networks' features are extracted within a receptive field [9]. The translation and rotation of fingerprints are due to cause problems in feature comparisons. To investigate the effect of translation, the metadata coupled to the fingerprints was used to create datasets with increasingly larger translations allowed.

The datasets consisted of matching and non-matching image pairs. Fingerprints were considered matching if their relative rotation was less than 45°C and the translation was below a number of pixels, see Table 1. Image pairs originating from the same finger but exceeding the thresholds for translation or rotation were not considered in the data sets. Non-matching image pairs are created by using images from other fingers and persons. Hence it is possible to create an extensive amount of non-matching data. The validation and test data were comprised of individuals not used to create the training data. Both in the case of creating matching and non-matching pairs.

Table 1. The number of matching image pairs in each dataset. The datasets are partitioned using the relative translation between fingerprint pairs. The training, validation and test datasets are constructed from mutually disjoint sets of enrolled fingers.

Translation (pixels)	Training	Validation	Test	Total
10	2795	253	199	3247
30	23677	2653	2807	29137
60	78622	9457	10263	98342
80	121292	14831	15939	152062
120	201625	25230	26404	253259

The metrics used to compare the different architectures were the equal error rate (EER), and the false negative matching rate (FNMR) at false matching rates (FMR) of 10^{-2}, 10^{-3}, and 10^{-4}, denoted FMR100, FMR1000, and FMR10000 [3].

3.2   Inference

In Table 2 the results of the inference on the test sets is presented. The architecture achieving the best results on all metrics is the Contrastive model. It is also possible to observe the vast difference between the Decision model and the other models, especially regarding the FMR metrics. By investigating Fig. 2, we can discern how changes in translation impact the ROC curve. Both the Contrastive and Triplet models have a substantial performance reduction when the allowed translation reaches 60 pixels.

In order to further examine why the Decision model is inferior to the two other models, the histogram of the matching scores is depicted in Fig. 3. The Triplet and Contrastive models have learned to differentiate between fingerprints by grouping them into two Gaussian distributions with some overlap. Whilst it is apparent that the Decision model is less flexible with its assessment of matching images. At closer inspection of the location of the non-matching examples in the Decision model, it is possible to see that for a few samples, it has placed a neer 100% certainty of them being matching. This is detrimental to the network's

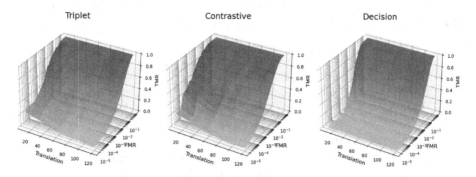

Fig. 2. Surface plots of ROC curve at a multitude of translations. TMR denotes the True Matching Rate, i.e. the ratio of correctly identified matching image pairs.

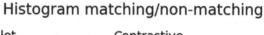

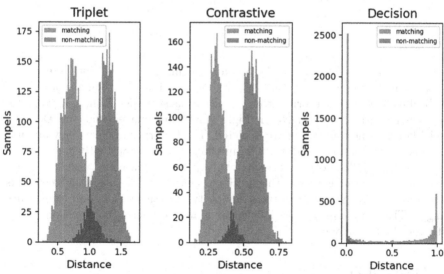

Fig. 3. Histogram showing the distribution of distances from the last layer of every model calculated on the test set with 30 pixels translation. This distance measure is the matching score between image pairs. The Decision model's distance can be interpreted as the probability of matching image pairs.

Table 2. Performance of the different network architectures evaluated on the test sets. It is notable that the Contrastive model outperforms the others on all metrics. The fingerprint images are collected from a sensor with size of [10 × 10] milimeters and 508 ppi.

Networks	Translation	Metrics			
		EER	FMR100	FMR1000	FMR10000
Triplet	10	0.06	0.21	0.54	0.69
	30	0.07	0.24	0.57	0.73
	60	0.10	0.42	0.72	0.86
	80	0.13	0.50	0.76	0.90
	120	0.17	0.63	0.85	0.93
Contrastive	10	0.04	0.12	0.34	0.54
	30	0.04	0.11	0.34	0.63
	60	0.06	0.21	0.57	0.82
	80	0.08	0.30	0.65	0.87
	120	0.11	0.46	0.76	0.92
Decision	10	0.09	0.55	0.93	0.98
	30	0.09	0.62	0.94	0.99
	60	0.12	0.71	0.95	0.99
	80	0.15	0.73	0.95	0.99
	120	0.20	0.78	0.96	0.99

performance at low FMR since these non-matching images will interfere with a lot of the matching examples that also get close to 100% matching score.

To illustrate the fingerprints that the networks struggles with, two examples are shown in Fig. 4 and Fig. 5. In Fig. 4, matching images that the Decision model labels as non-matching are depicted. The fingerprints have a low amount of translation but missing data. The missing data is persistently hard for all networks to handle, both in the matching and non-matching cases. Figure 5 shows a non-matching pair with additional characteristics which all networks struggle with, namely global pattern similarity, over saturation, and low-quality images. These fingerprints are labeled as matching using the Decision model with very high confidence.

4   Discussion

Based on the experiment results, it can be concluded that the Siamese model with contrastive loss outperformed the other models in terms of recognition accuracy, achieving an EER between 4% to 11%, which is approximately 6.5% points lower than the categorical schemed model. Meanwhile, the categorical scheme performed poorly, especially when recognizing images with high confidence. Additionally, the networks were mostly impacted by translation when it reached one-fourth of the image.

The results also indicate that the choice of neural network architecture has a significant impact on the performance of image recognition systems. Specifically, the Siamese models based on contrastive loss and triplet loss, which create feature embeddings in a Euclidean space, exhibited superior performance in terms of recognizing images with high confidence. As shown in Fig. 3, the distance histograms of these models had a Gaussian distribution, indicating a softer decision threshold. In contrast, the Decision model relied on binary classification, making it more challenging to establish a threshold that balances false matching rates with true positive rates, as shown in Table 2. Although the Decision model performed similarly to other models in the EER metric, its binary classification attribute was apparent, highlighting the importance of selecting an appropriate architecture for authentication systems based on neural networks. These findings suggest that clustering-based architectures, such as the Siamese models, are advantageous in designing such systems.

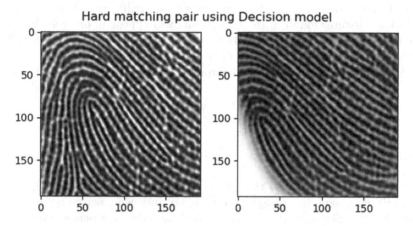

Fig. 4. Pair of matching fingerprints. The Decision model struggle with recognizing the similarities due to the missing fingerprint data, presumably caused by the finger not covering the whole sensor. The matching score is close to 0.

As described in the Sect. 3, the networks direct considerable attention to the global structure of the fingerprint when determining the similarity between image pairs, see Fig. 4 and Fig. 5. Ridge patterns, as well as missing data and saturation, cause miss classified fingerprints. The behavior of the networks in finding global structures for matching is also reported by [17] and in their preceding work [15]. Both of which rely in part upon Triplet networks. Even though all networks are affected by imposter pairs with similar global patterns, the Decision model performs considerably worse on all metrics, see Table 2. For future work, the difference in features between the models would be an interesting subject to pursue to establish the reason behind the performance discrepancy further.

Hard non-matching pair using Decision model

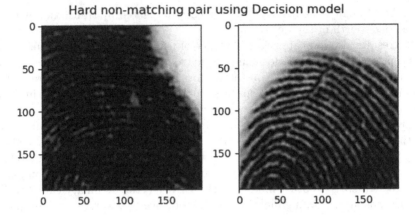

Fig. 5. Example of two non-matching images that would be considered matching by the network. Matching score greater than 0.999.

All the networks are susceptible to performance reduction when varying the translation. Features found in the fingerprints will appear in different locations of the feature maps when fingerprints are translated relative to each other. This poses a problem for all the networks as the last layers will need to determine the relative location of features in the fingerprints to make accurate predictions. As the translation increases, this becomes harder, which is shown in Fig. 2. Here, the Contrastive and Triplet models have a significant change in their performance when going from 30 pixels to 60 pixels translation.

5   Conclusion

In this work, we have studied the impact of the choice of network architecture when performing one-shot recognition of fingerprints using neural networks. We showed that the Siamese network based on Contrastive loss performed the best on all test sets. We also recognized that the Decision model had difficulties performing well at low false matching rates, due to granting very high certainty of matching to some non-matching fingerprints. We have also investigated the impact of relative translation between fingerprints. The models handling the translation best, saw a significant drop in performance when going from a dataset with an allowed translation of 30 pixels to one with 60 pixels.

References

1. Alrashidi, A., Alotaibi, A., Hussain, M., AlShehri, H., AboAlSamh, H.A., Bebis, G.: Cross-sensor fingerprint matching using siamese network and adversarial learning. Sensors **21**(11) (2021). https://doi.org/10.3390/s21113657
2. Anand, V., Kanhangad, V.: PoreNet: CNN-based pore descriptor for high-resolution fingerprint recognition. IEEE Sens. J. **20**(16), 9305–9313 (2020). https://doi.org/10.1109/JSEN.2020.2987287

3. Cappelli, R., Ferrara, M., Maltoni, D., Turroni, F.: Fingerprint verification competition at ijcb2011. In: 2011 International Joint Conference on Biometrics (IJCB), pp. 1–6 (2011). https://doi.org/10.1109/IJCB.2011.6117488
4. Chowdhury, A., Kirchgasser, S., Uhl, A., Ross, A.: Can a CNN automatically learn the significance of minutiae points for fingerprint matching? In: 2020 IEEE Winter Conference on Applications of Computer Vision (WACV), pp. 340–348 (2020). https://doi.org/10.1109/WACV45572.2020.9093301
5. Hadsell, R., Chopra, S., LeCun, Y.: Dimensionality reduction by learning an invariant mapping. In: 2006 IEEE Computer Society Conference on Computer Vision and Pattern Recognition (CVPR'06), vol. 2, pp. 1735–1742 (2006). https://doi.org/10.1109/CVPR.2006.100
6. Lee, H.C., Gaensslen, R.E.: Advances in Fingerprint Technology, 2nd edn. CRC Press Inc (2001)
7. He, K., Zhang, X., Ren, S., Sun, J.: Deep residual learning for image recognition. In: 2016 IEEE Conference on Computer Vision and Pattern Recognition (CVPR), pp. 770–778 (2016). https://doi.org/10.1109/CVPR.2016.90
8. Lam, T., Nilsson, S.: Application of convolutional neural networks for fingerprint recognition (2018), student Paper
9. Luo, W., Li, Y., Urtasun, R., Zemel, R.: Understanding the effective receptive field in deep convolutional neural networks. In: Proceedings of the 30th International Conference on Neural Information Processing Systems, pp. 4905–4913. NIPS'16, Curran Associates Inc., Red Hook, NY, USA (2016). https://doi.org/10.5555/3157382.3157645
10. Salomon, G., Britto, A., Vareto, R.H., Schwartz, W.R., Menotti, D.: Open-set face recognition for small galleries using siamese networks. In: 2020 International Conference on Systems, Signals and Image Processing (IWSSIP), pp. 161–166 (2020). https://doi.org/10.1109/IWSSIP48289.2020.9145245
11. Schroff, F., Kalenichenko, D., Philbin, J.: FaceNet: a unified embedding for face recognition and clustering. In: 2015 IEEE Conference on Computer Vision and Pattern Recognition (CVPR) (2015). https://doi.org/10.1109/cvpr.2015.7298682
12. Snoek, J., Larochelle, H., Adams, R.P.: Practical Bayesian optimization of machine learning algorithms. In: Proceedings of the 25th International Conference on Neural Information Processing Systems - Volume 2, pp. 2951–2959. NIPS'12, Curran Associates Inc., Red Hook, NY, USA (2012). https://doi.org/10.5555/2999325.2999464
13. Tan, M., Le, Q.V.: EfficientNet: rethinking model scaling for convolutional neural networks. CoRR abs/1905.11946 (2019). https://arxiv.org/abs/1905.11946
14. Zhang, D., Liu, F., Zhao, Q., Lu, G., Luo, N.: Selecting a reference high resolution for fingerprint recognition using minutiae and pores. IEEE Trans. Instrum. Meas. 60(3), 863–871 (2011). https://doi.org/10.1109/TIM.2010.2062610
15. Zhang, F., Feng, J.: High-resolution mobile fingerprint matching via deep joint KNN-triplet embedding. In: Proceedings of the AAAI Conference on Artificial Intelligence, vol. 31, issue 1 (2017). https://doi.org/10.1609/aaai.v31i1.11088
16. Zhang, F., Xin, S., Feng, J.: Deep dense multi-level feature for partial high-resolution fingerprint matching. In: 2017 IEEE International Joint Conference on Biometrics (IJCB), pp. 397–405 (2017). https://doi.org/10.1109/BTAS.2017.8272723

17. Zhang, F., Xin, S., Feng, J.: Combining global and minutia deep features for partial high-resolution fingerprint matching. Pattern Recogn. Lett. **119**, 139–147 (2019). https://doi.org/10.1016/j.patrec.2017.09.014
18. Zhu, L., Xu, P., Zhong, C.: Siamese network based on CNN for fingerprint recognition. In: 2021 IEEE International Conference on Computer Science, Electronic Information Engineering and Intelligent Control Technology (CEI), pp. 303–306 (2021). https://doi.org/10.1109/CEI52496.2021.9574487

3D Facial Reconstruction Based on a Single Image Using CNN

Ramzi Agaba[1]([✉])(iD), Mehdi Malah[2](iD), Fayçal Abbas[3](iD),
and Mohamed Chaouki Babahenini[4](iD)

[1] ReLaCS2 Laboratory, Computer Science Department, University Larbi Ben Mhidi,
Oum El Bouaghi, Algeria
ramzi.agaba@univ-oeb.dz

[2] ICOSI Laboratory, Computer Science Department, University of Abbes Laghrour
Khenchela, Khenchela, Algeria
malah.mehdi@univ-khenchela.dz

[3] LESIA Laboratory, Computer Science Department, University of Abbes Laghrour
Khenchela, Khenchela, Algeria
abbas_faycal@univ-khenchela.dz

[4] LESIA Laboratory, Computer Science Department, University of Mohamed Khider
Biskra, Biskra, Algeria
mc.babahenini@univ-biskra.dz

Abstract. In recent years, 3D facial reconstruction marked its presence in several areas, such as biometric security, computer vision, image synthesis, and video games. Many facial recognition applications require accurate 3D reconstruction; however, this task is complex and requires many calculations. This paper presents a new method based on CNN, which offers an automatic solution for the 3D reconstruction of the face depending on a single 2D image. Our method operates in three steps: the first step is to train a convolutional neural network with a self-made dataset to predict and detect the landmarks of the face and estimate the positions accurately from a single facial image in the image space. The second step involves producing the face's geometric shape (mesh). Finally, the third step is to do an automatic translation between the 3D space of the object and the 2D image space to determine the texture referrals that correspond to each face polygon; thus, our method produces superior results in terms of accuracy and visual quality, considering all the parameters of the model (shape, expression, reflectance, and lighting) as inputs. Our method is simple, easy to implement, and offers a real-time 3D reconstruction of the face.

Keywords: Deep learning · 3D reconstruction · 3D facial
reconstruction · Convolutional neural network · CNN

1 Introduction

Three-dimensional face reconstruction pushed the limits of computer vision further, especially in several areas such as augmented reality, computer security, and

A. Bennour et al. (Eds.): ISPR 2023, CCIS 1940, pp. 15–26, 2024.
https://doi.org/10.1007/978-3-031-46335-8_2

computer graphics. The improvements applied to GPUs offer and ease the automatic 3D reconstruction of faces in real-time. The traditional techniques used to extract facial features are less stable regarding the localization of objects in an image. The complexity imposed by the three-dimensional facial reconstruction lies in the variation of illumination and position and the problems of occlusion and expression. Recent approaches based on deep learning models have emerged, especially convolutional neural networks (CNNs), which have marked their presence in several areas of computer vision. This paper presents a method capable of reconstructing a face from a single 2D image. Our method operates in three parts: the first consists of training a CNN on our dataset to match the landmarks in the image space, and then a face mesh is reconstructed. Finally, we perform a translation between the 3D space of the object and the 2D image space (texture) to determine the coordinates of the textures which correspond to each face polygon. Our paper's contributions are as follows:

- We propose a new CNN-based method to construct a 3D reconstructed model of the human face from a single image.
- We introduce a sizable dataset of facial images with their 3D model that was used to train and evaluate our model.
- We propose a real-time reconstruction model for facial geometry which offers a suitable generalization of the facial points.

2   Related Work

The reconstruction of a 3D face from a single 2D image is an essential topic in computer vision. However, recent approaches based on learning, in which unique images form models, have yielded promising results for 3D monocular face reconstruction. Nevertheless, they need help with an incorrectly posed face and depth ambiguity. As a consequence, methods based on a single image can generally only provide an approximation of facial geometry. Several learning-based methods can perform 3D facial reconstructions, which are also founded on a single image. Each of the methods has its advantages. These methods are primarily different regarding the time required to generate the desired result and the platform on which the reconstruction is executed.

[3] They proposed a new approach for 3D reconstruction of the face (Deep3D) using CNN; they also proposed their hybrid-level loss function, which was based on many characteristics like the identity, expression, texture and pose; it also performs facial reconstruction by exploiting additional information from different angles to reconstruct 3D faces more accurately. Their approach is fast and accurate, surpassing prior methods in several terms, such as accuracy and robustness.

[15] Their approach, called 3D Dense Face Alignment (3DDFA), was based on Cascaded CNNs, in which they based their work to provide results even with faces in many poses at different degree alignments. They used 3D information to synthesize facial data into profile views to deliver abundant cases for training.

Their approach provided good and more accurate results surpassing other state-of-the-art methods.

[4] They proposed a method that concurrently reconstructs 3D facial features and offers dense alignments. They developed a 2D representation, a "card of UV position", representing a complete 3D face in the UV space. They used a CNN to predict it from a single 2D picture. Meanwhile, experiments on different datasets show that this method improves 3D reconstruction and alignment tasks over others from afar. They also proved that their method works faster and is suitable for real-time use.

[11] They introduced a CNN frame that continuously extracts the shape of the picture. Two major blocks form the proposed architecture: a network that recovers the face's gross geometries (GrosseNet), followed by a CNN that combines the face characteristics of this geometry (FineNet). The proposed networks are linked to a new layer that converts a given depth picture into a 3D model. Consequently, their development begins with a monitored phase and uses synthetic images, followed by an unmonitored phase that only uses available face images. Their demonstration and comparisons show that their method surpasses recent approaches, primarily if used for 3D recognition tasks, since it provides the closest 3D models to the images.

[12] They proposed an auto-supervised architecture that provides reliable constraints on the position of the face and depth estimation. Additionally, three new loss functions for multi-view coherence have been proposed, pixel loss, depth loss, and epipolar loss of face points for reperception. Extensive research on facial alignment and 3D facial reconstruction has demonstrated the method's superiority over existing methods. Their approach has been accurate and robust, mainly when dealing with significant variations in expressions, poses, and illuminance. Compared to other methods, the results show that their MGCNet can attain excellent results even without fancy-marked data.

[10] By carefully designing the new Siamese CNN (SCNN). They proposed a complete resolution to the problem of 3D reconstruction, using contrast loss to increase the interclass distance and decrease the intraclass distance for the 3D Morphable face Model (3DMM) output parameters for the same individual in the same class's (3DMM) settings. They have also proposed an identity loss function to keep the identity information in the space of features for the same individual. Their Model could learn different representations that are more discriminatory for the identity and could generalize for other pose variants. Experiments on different datasets like the 300W-LP and AFLW2000-3D showed their method's effectiveness by comparing with the state-of-the-art.

3   Method

The planning of our method is made up of three steps. The first step is to perform training on a dataset that we have created using a CNN whose input is a 2D image, while the output is a vector including the face landmarks. The second step is to generate the 3D mesh, represented by 468 vertexes and 880 faces.

Lastly, we go deep into the texture mapping phase, which consists of plating the texture on the 3D model.

3.1 Creating the Dataset

Setting up a dataset of images with their resulting 3D objects took work regarding the effect it opposes to the results. We collected and selected the data based on many characteristics, transformations, and selection phases.

Facial Images Collection. At first, we started by collecting facial images from different sources HELEN [9] and some particular photographs like the family and friends portraits to increase the data size, each of the latter covering different human face positions. About 20000 images were collected from these sources, and the size of the images varies between (1024×1024) and (224×224) pixels. Then, using a script with Haar Cascade's classifier [13] to find, detect and crop human faces from each image and then resize each of them to (224×224). After that, we moved into a selection phase; we had to look closely and manually select the images that best fit our needs. We attained to considerable data size, which was around 12000 images.

(Fig. 1) shows an example of the different stages of detection and selection phases.

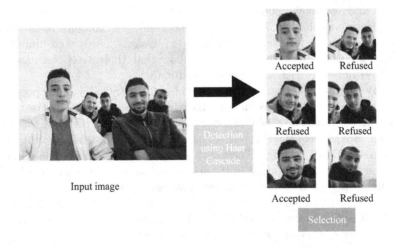

Fig. 1. Different stages for the detection and selection phases.

Face Detection. In this step, we passed each of the 12 thousand images to Facemesh [6], which allowed us to detect different human face landmarks. We used the latter due to the significant variability and importance of the resulting landmarks in the perception and awareness of the human face. They do grant a construction of a plausible and smooth surface representation according to the Catmull Clark subdivision [2].

The topology incorporates 468 points (x, y, z) (face landmarks) and 880 triangulation. As we know, estimating the depth of pixels in a 2D image is still a challenge. At the end of this step, we eliminated the photos poorly detected with Facemesh; up to 83% of the images had suitable predictions. Eventually, 10K images have been noted as pleasing results (Fig. 2).

Points without Triangulation   Points with Triangulation

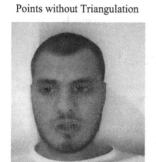 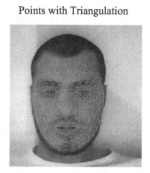

Fig. 2. Results of the detection phase.

Selection of 3D Objects. From the results of the previous steps, An algorithm whose purpose is creating a corresponding 3D object (geometry) for each image, using the x, y, and z coordinates of the cue points and the triangulation, was executed. The Texture mapping was presented in a UV vector space. We had to perform another manual selection task to ensure the objects' accuracy, cleanliness, and clarity. We ended up with 8000 instances that provided clear and clean results fitting our demands. Each coordinate has been saved in a format that can be directly represented on the texture scale (image space), where each coordinate will have a value between 0 and 1. However, to switch to a 3D representation, a translation between 2D and 3D spaces must be present (Fig. 3).

Accepted   Refused

Fig. 3. Selection phase of 3D objects.

Dataset Preprocessing. This step is essential to enhance the performance and speed of the training; it is accomplished before the training task. In this model, we inherently applied two pretreatment processes.

- Transforming the dataset images (inputs) into a 4-dimensional array (See Fig. 4).
- Transformed the 3D objects (outputs) into a 2-dimensional array (See Fig. 5).

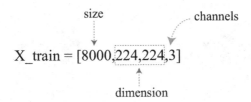

Fig. 4. Input shape of our CNN architecture.

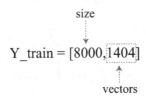

Fig. 5. Output shape of our CNN architecture.

The 1404 vectors are demonstrated as follows:
$$[x_0, x_1, x_2...x_{468}, y_0, y_1, y_2...y_{468}, z_0, z_1, z_2...z_{468}]$$
After the preprocessing phase, 75% of the images (6000 instances) were intended for the training phase and 25% (2000 instances) for the test phase.

3.2 CNN Architecture

CNNs are layer-based architectures that will endure various procedures and operations to extract the essential characteristics from the images. These characteristics will be transmitted to a fully connected neural network to perform the learning phase. As we pass via the network, each convolution layer is pursued by an activation layer (RELU).

Our model's architecture contains six convolution layers and five max-pooling layers. (Fig. 6) represents a brief and non-detailed architecture.

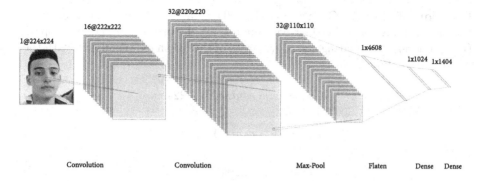

Fig. 6. Our non-detailed CNN architecture.

Table 1 summarizes the different layers of our convolutional network.

Table 1. Description of our CNN architecture.

Layer	Description	Input	Output
1	Conv (3×3) RELU activation	$224 \times 224 \times 3$	$222 \times 222 \times 16$
2	Conv (3×3) RELU activation	$222 \times 222 \times 16$	$220 \times 220 \times 32$
3	Max_Pool ($2 \times 2, stride2$)	$220 \times 220 \times 32$	$110 \times 110 \times 32$
4	Conv (3×3) RELU activation	$110 \times 110 \times 32$	$108 \times 108 \times 64$
5	Max_Pool ($2 \times 2, stride2$)	$108 \times 108 \times 64$	$54 \times 54 \times 64$
6	Conv (3×3) RELU activation	$54 \times 54 \times 64$	$52 \times 52 \times 128$
7	BatchNormalization()	$52 \times 52 \times 128$	$52 \times 52 \times 128$
8	Max_Pool($2 \times 2, stride2$)	$52 \times 52 \times 128$	$26 \times 26 \times 128$
9	DropOut()	$26 \times 26 \times 128$	$26 \times 26 \times 128$
10	Conv (3×3) RELU activation	$26 \times 26 \times 128$	$22 \times 22 \times 256$
11	Max_Pool($2 \times 2, stride2$)	$22 \times 22 \times 256$	$11 \times 11 \times 256$
12	Conv (3×3) RELU activation	$11 \times 11 \times 256$	$7 \times 7 \times 512$
13	BatchNormalization()	$7 \times 7 \times 512$	$7 \times 7 \times 512$
14	Max_Pool($2 \times 2, stride2$)	$7 \times 7 \times 512$	$3 \times 3 \times 512$
15	DropOut ()	$3 \times 3 \times 512$	$3 \times 3 \times 512$
16	Flatten()	$3 \times 3 \times 512$	4608
17	Dense()	4608	1024
18	BatchNormalization()	1024	1024
19	DropOut()	1024	1024
20	Dense()	1024	1404

3.3 3D Reconstruction

At last, we are using the model to acquire the different face landmarks, which will be employed to create the final 3D geometry.

Creation of the 3D Geometry. The 3D points of the face obtained from the model, represented by their coordinates, are defined in the 2D image space (texture level). Calculating the coordinates of the 3D shape is supplied with the mathematical functions:

$$F(x) = (x - centreX) \times 224 \tag{1}$$

$$G(y) = (y - centreY) \times 224 \tag{2}$$

$$H(z) = (z \times 200) - 100 \tag{3}$$

– x, y, and z are the prediction coordinates.
– We chose the nose as a center point (centreX, centreY).

The coordinates of the 468 points were assembled and represented on the 3D stage, including the triangulation; The result is a 3D geometric shape of the face while still without the texture mapping (Fig. 7).

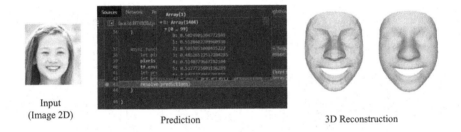

| Input (Image 2D) | Prediction | 3D Reconstruction |

Fig. 7. Example of the 3D reconstruction from a 2D image (without texture).

Texture Plating. The texture plating step consists of texturing all triangles of the resulting object with the corresponding UV mapping obtained with the model (Fig. 8).

Fig. 8. Example of the 3D reconstruction from a 2D image (with texture).

4   Experimental

4.1   Detailed Experimental

We used the constructed dataset to execute the training step of the CNN model. However, we reduced the input image size to 244×244 to decrease the computations and keep respect for the quality and detail of the latter. We also employed the Adam optimizer [7] with an initial leaning rate equal to 1e-3, $\beta 1 = 0.9$, $\beta 2 = 0.999$, and a batch size equivalent to 64. The training phase of the model was accomplished using a GPU cloud (from Lambda Labs [8]), whose main graphics card (GPU): is $4 \times$ NVIDIA Pascal Based GPUs (11 GB).

4.2   Comparison

To verify the effectiveness of our approach, we compare our solution with two recent methods [3,15] by accomplishing a quantitative and qualitative evaluation by comparing the geometry obtained from each of the approaches with those called references (Florence [1], AFLW2000-3D [14]).

Quantitative Comparison. We used the Hausdorff distance [5] as a comparison metric. This metric qualified to calculate the error between two shapes as the distance between all the points between both geometries, the so-called ground truth, and the geometry generated from each method; the following formula calculates the Hausdorff distance:

$$d_H(X,Y) = \max(h(X,Y), h(Y,X)) \tag{4}$$

$$h(X,Y) = \max_{x \in X} \min_{y \in Y} \|x - y\| \tag{5}$$

$\|.\|$ is an underlying norm on the points of X and Y.

Table 2 compares our method and the two state-of-the-art approaches [3,15] on two datasets employed as reference entities.

Table 2. Quantitative comparison.

Methods	Florence	AFLW2000-3D
Deep3D	RMS : 0.039038	RMS : 0.018764
3DDFA	RMS : 0.035994	RMS : 0.006466
Ours	**RMS : 0.016803**	**RMS : 0.021883**

Our method submits a better approximation of the resulting objects as it offers a reconstruction model with an error of 0.0168 on the Florence dataset, demonstrating its effectiveness for 3D face reconstruction.

Qualitative Comparison. (Figs. 9, and 10) presents a qualitative comparison of the geometric shapes generated with our approach and different state-of-the-art approaches [3, 15].

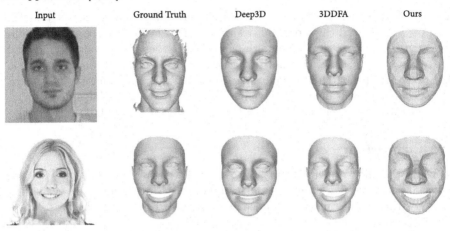

Fig. 9. Qualitative comparison of geometric shapes without textures generated by our method and [3, 15].

Our approach presents a good approximation of the generated mesh.

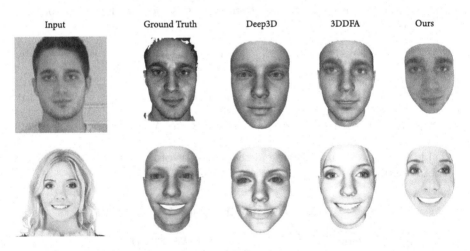

Fig. 10. Qualitative comparison of geometric shapes with textures generated by our method and [3, 15].

(Figures 9, and 10) shows that our method obtains good estimations of all model parameters, i.e., the scene's pose, shape, expression, reflectance, and lighting. Comparing the results of all approaches demonstrates that our approach also obtains a good imitation of the reflection and illumination of the colored skin.

5   Conclusion

Our method was based on the use CNNs, which generate a 3D reconstruction of the face from a single 2D image. Our approach produces superior accuracy and visual quality results, demonstrated with experiments on the two facial datasets, AFLW2000-3D and Florence. That illustrates the effectiveness of our approach for 3D facial reconstruction by comparing it with other state-of-the-art methods. Our approach is also simple to implement, offering a real-time 3D reconstruction of the face.

References

1. Bagdanov, A.D., Del Bimbo, A., Masi, I.: The florence 2D/3D hybrid face dataset. In: Proceedings of the 2011 Joint ACM Workshop on Human Gesture and Behavior Understanding, pp. 79–80 (2011)
2. Catmull, E., Clark, J.: Recursively generated b-spline surfaces on arbitrary topological meshes. Comput. Aided Des. **10**(6), 350–355 (1978)
3. Deng, Y., Yang, J., Xu, S., Chen, D., Jia, Y., Tong, X.: Accurate 3D face reconstruction with weakly-supervised learning: From single image to image set. In: Proceedings of the IEEE/CVF Conference on Computer Vision and Pattern Recognition Workshops (2019)
4. Feng, Y., Wu, F., Shao, X., Wang, Y., Zhou, X.: Joint 3D face reconstruction and dense alignment with position map regression network. In: Proceedings of the European Conference on Computer Vision (ECCV), pp. 534–551 (2018)
5. Huttenlocher, D.P., Klanderman, G.A., Rucklidge, W.J.: Comparing images using the Hausdorff distance. IEEE Trans. Pattern Anal. Mach. Intell. **15**(9), 850–863 (1993)
6. Kartynnik, Y., Ablavatski, A., Grishchenko, I., Grundmann, M.: Real-time facial surface geometry from monocular video on mobile GPUs. arXiv preprint arXiv:1907.06724 (2019)
7. Kingma, D.P., Ba, J.: Adam: a method for stochastic optimization. arXiv preprint arXiv:1412.6980 (2014)
8. LambdaLabs: GPU cloud, workstations, servers, laptops for deep learning. https://lambdalabs.com/
9. Le, V., Brandt, J., Lin, Z., Bourdev, L., Huang, T.S.: Interactive facial feature localization. In: Fitzgibbon, A., Lazebnik, S., Perona, P., Sato, Y., Schmid, C. (eds.) ECCV 2012. LNCS, vol. 7574, pp. 679–692. Springer, Heidelberg (2012). https://doi.org/10.1007/978-3-642-33712-3_49
10. Luo, Y., Tu, X., Xie, M.: Learning robust 3D face reconstruction and discriminative identity representation. In: 2019 IEEE 2nd International Conference on Information Communication and Signal Processing (ICICSP), pp. 317–321. IEEE (2019)
11. Richardson, E., Sela, M., Or-El, R., Kimmel, R.: Learning detailed face reconstruction from a single image. In: Proceedings of the IEEE Conference on Computer Vision and Pattern Recognition, pp. 1259–1268 (2017)
12. Shang, J., et al.: Self-supervised monocular 3D face reconstruction by occlusion-aware multi-view geometry consistency. In: Vedaldi, A., Bischof, H., Brox, T., Frahm, J.-M. (eds.) ECCV 2020. LNCS, vol. 12360, pp. 53–70. Springer, Cham (2020). https://doi.org/10.1007/978-3-030-58555-6_4

13. Viola, P., Jones, M.: Rapid object detection using a boosted cascade of simple features. In: Proceedings of the 2001 IEEE CSociety Conference on Computer Vision and Pattern Recognition CVPR 2001, vol. 1, p. I. IEEE (2001)
14. Zhu, X., Lei, Z., Liu, X., Shi, H., Li, S.Z.: Face alignment across large poses: a 3D solution. In: Proceedings of the IEEE Conference on Computer Vision and Pattern Recognition, pp. 146–155 (2016)
15. Zhu, X., Liu, X., Lei, Z., Li, S.Z.: Face alignment in full pose range: a 3D total solution. IEEE Trans. Pattern Anal. Mach. Intell. $41(1)$, 78–92 (2017)

Towards the Detectability of Image Steganography Using NNPRTOOL (Neural Network for Pattern Recognition)

Ayidh Alharbi[✉]

School of Computer Science, University of Tabuk, Tabuk, Saudi Arabia
a.alhunaini@ut.edu.sa

Abstract. The term steganography represent the science of hiding secret information in a carrier medium like Image, Audio, Video, etc. This field has been participated in multidisciplinary applications and different sectors. For instance, in Healthcare, E-commerce, Access control, and Databases. According to literature, some of those applications, and their analytical, classification data and results are studied. This include current challenges of detectability which can be done with or without using a machine learning classifier. In this research, an additional experiment of Neural Network Pattern Recognition (NNPRTOOL) is applied to classify original and stego images belong to two main classes and directories. This experiment contributes in the robustness of steganography in general. Moreover, it highlights the evolution of image steganalysis research using a machine learning classifier. However, the potential of such tools like NNPRTOOL, Support Vector Machine (SVM), and ENSEMBLE Classifier have shown significant results in image classifications.

Keywords: Image · Steganography · Neural Network · SVM · NNPRTOOL · MATLAB

1 Introduction

Steganalysis aims to detect the existence of secret information in an image. It is either active or passive. If passive steganalysis aim to find the existences of secret information. Active steganalysis aims to find the secret message its self [1]. In digital steganography, Images represent the most digital carriers medium [2]. However, in this research, the potential of machine learning classifier are studied to distinguish original and stego images through NNPRTOOL. The argument here if the machine learning classifier can recognise and learn to differ a bunch of experimental images. In this context, a good survey examines image classification and its current practice and prospects in [3]. However, Deep neural network is an emerging machine learning method that has proven its potential for different classification tasks [4]. The evolution of Convolutional Neural Network (CNN) is a form of feedforward neural network with a deep learning structure and convolution computations known between 1980 and 1990 [5]. Moreover,

© The Author(s), under exclusive license to Springer Nature Switzerland AG 2024
A. Bennour et al. (Eds.): ISPR 2023, CCIS 1940, pp. 27–37, 2024.
https://doi.org/10.1007/978-3-031-46335-8_3

the Convolutional Neural Network (CNN)-based deep learning has become more popular in learning the features from image directories. For instance, SVM is also a high-performance classification algorithm widely used in image classification. The image classification based deep learning participates in many applications such as: Medical imaging system, Copyright control, Automatic driving, Image caption generation, And computer vision applications [6]. In such problem of classification, extracting the patterns and features from the images is needed as pre-step for learning. Simultaneously, the pattern recognition is playing a role for training the neural network to assign the correct target classes to a set of input patterns. This include many applications like Image Processing, Speech Recognition, and Text Classification. More examples of its application highlighted in [7]. By Considering the feature extraction the key of such algorithms. Main example of feature extraction techniques used on images are, Histogram of oriented gradients (HOG), Scale-invariant feature transform (SIFT), Content-Based Image Retrieval (CBIR), and Local Binary Patterns (LBP). In this research, the local binary patterns is used to extract a descriptive feature from all images. Therefore, there are a common steps of image classifications in NNPRtool that are mentioned in Fig. 1.

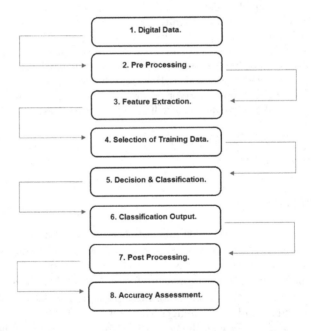

Fig. 1. Steps for image classification according to Mehat, 2016.

Local Binary Pattern (LBP): Is a simple yet very efficient texture operator (Descriptor) which labels the pixels of an image by thresholding the neighbourhood of each pixel and considers the result as a binary number. The idea of

LBP is motivated by the fact that some binary patterns occur in a more commonly texture of images than others. A local binary pattern is called uniform if the binary patterns contain at most two 0–1 or 1–0 transitions. For instance, 00010000 (2 transitions) is a uniform pattern, 01010100 (6 transitions) is not. The function (extractLBPFeatures) extracts 59 identical lbp's in any image. An example of 4 images lbp's in Fig. 2.

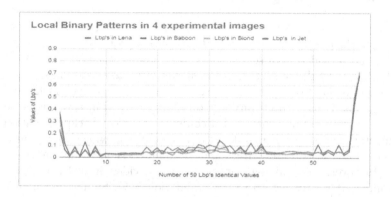

Fig. 2. Lbp's Values in the proposed images.

2   Related Work

In the field of machine learning, there are many studies have been participating in a wide range of image classification and recognition. From the impact of machine learning perspectives, NNprtool help to study different Identification, classification, and Recogniton applications. In this context, A comparison between the traditional machine learning and deep learning image classification algorithms are studied in [8]. The potentials of Deep learning and Neural Network have been participated in many computer vision and image classification fields. Several example of texture classification, Pattern recognition, Application of object detection were applied in [6]. Moreover, An experiment of using convolutional neural network for images classification in [9]. Another example of using deep learning and pre-trained convolutional neural network models to classify set of images in [10]. In [11], the discussion of artificial neural network potentials to classify handwritten digit recognition. In [12], it addresses the using of Convolutional neural network to classify many different medical imaging applications. In [13], the Deep Neural Network analysis and performance of popular images datasets: Alex Net, GoogleNet, and ResNet50 are provided for identifying objects in Image and Videos. Indeed [14] provides a survey of SVM and their application in image classification. Moreover, In [15] using MATLAB/NPRTOOL environment, a two-layer feed-forward network with sigmoid output neurons can be trained in order to classify Road Roughness Conditions. Another study in [16] represent a comprehensive description of the current technology of deep learning and its potentials in image classification and health care sector. In conclusion, these researches have been done with experimental images in several image datasets, Application, Image classification and Pattern recognition field.

3   Benchmark Image Datasets Experiments

The potential of Neural Network Pattern Recognition (NNPR tool) is very helpful, useful to the image classification and recognition problem. It provides several examples and provided data sets such as: Iris Flowers, type of glass, Wine vintage, etc. It assigns the correct target classes in order to classify input patterns. In this experiment, two layers feed-forward network with sigmoid hidden and output neurons is used to classify a total of 164 images in two classes (Clean and Stego images) trained with scaled conjugate gradient backpropagation. All experimental images have been taken from two datasets which are Signal Image Processing Instituete dataset (SIPI) and Computer Vision Group image dataset (CVG). However, the training parameters are divided into 3 samples which are: Training 70%, Testing 15%, and Validation 15%. Similar example found in [17,18]. Each directory consists of 82 images. However, the benchmark images classification experiment through NNPRTOOL are given in Fig. 3.

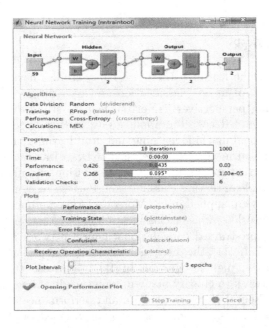

Fig. 3. Neural network training overview.

In Fig. 3, the tool (nntraintool) consist of 4 main labels which are: Neural network and its main description, set of algorithms used for the classification process, its progress and performance, And the type of its important plots. However, the clean and stego images classification process is motivated by the same examples. Another related work has been done in [19]. Based on Fig. 3, the main important plots are: plotperform, plottrainstate, ploterrhist, plotconfusion, and plotroc. An example of their functionalities as follows:

1. **Performance: (plotperform)**
   plotperform(trainingRecord) plots the training record through the function TRAIN or a specific training function. Here a feed-forward network is used for training simplefit-dataset:
   [x,t] = simplefit-dataset.
   net = feedforwardnet(10).
   $\big[$net,tr$\big]$ = train(net,x,t).
   plotperform(tr) [6].
2. **Training State (plottrainstate)**
   plottrainstate stands for plot training state values. plottrainstate(training-Record) plots the training states returned by the function TRAIN or a specific training function [6].
3. **Error Histogram (ploterrhist)**
   ploterrhist(errors) takes error data and plots a histogram.
   ploterrhist(errors1'name1', errors2, name2, ...) plots multiple error sets.
   ploterrhist(..., 'bins',bins) plots using an optional parameter that overrides the default number of error bins (20).
   Here a feed-forward network is used to solve a simple fitting problem:
   [x,t] = simplefit-dataset.
   net = feedforwardnet(10).
   net = train(net,x,t).
   y = net(x).
   e = t - y.
   ploterrhist(e, 'bins', 30) [6].
4. **Confusion Matrix (plotconfusion)**
   Another plot is plotconfusion, it takes the target and output data to generates a confusion plot. However, the target data represent the ground truth labels in 1-of-N form (in each column, a single element is 1 to indicate the correct class, and all other elements are 0). The output data are derived from a neural network that performs classification. It can either be in 1-of-N form, or may also be probabilities where each column sums to 1.
   plotconfusion generates several confusion plots in one figure. It accumulates (targets1, outputs1, 'name1', targets2, outputs2, 'name2',...), and prefixes the character strings specified by the 'name' arguments to the titles of the appropriate plots. In the following, an example of how to train a pattern recognition network and plot its accuracy.
   [x,t] = simpleclass-dataset.
   net = patternnet(10).

net = train(net,x,t).
y = net(x).
plotconfusion(t,y) [6].

5. **Receiver Operating Characteristics (plotroc)**
   Aother plot is plotroc, it takes target and output data in 1-of-N form where (each column vector is all zeros with a single 1 indicating the class number), and output data and generates a receiver operating characteristic plot. The ROC is generated by the threshold between the rate between (TPR) and (FPR) which are True positive and False Positive Rate respectively. For example, in Iris data set, [tpr,fpr,thresholds] = roc(irisTargets,irisOutputs). In the ROC plots, if the receiver operating line hugging the left and top sides of axis, it indicates a better classification.
   plotroc(targets, 1, outputs1, 'name1', targets2, outputs2, names2,...) generates a variable number of confusion plots in one figure. Here a pattern recognition network is trained and its accuracy plotted:
   [x,t] = simpleclass-dataset.
   net = patternnet(10).
   net = train(net,x,t).
   y = net(x).
   plotroc(t,y) [6].

Steps for Images training and classification model

1. Get the file path from each directory (Original, Stego).
2. Read all images from the two file paths (Original and stego).
3. If the image in RGB colour, convert all of them to Gray- scale.
4. From all 164 images, extract the local binary patterns from each image using extractLBPFeatures. This will produce 59 Identical vectors from each image.
5. Gather all Lbp's in an index of lbp's features matrix. The size of this matrix is (59 × 164).
6. Specify the target of each class based on the transition values of the uniform local binary patterns. For instance, stego target class transition is [0,1]. And original target class transition is [1,0].
7. Start training the model using NPR tool with 2 hidden layers.
8. Out of all training parameters, auto split the training model parameters in a random sample divisions which are 70% for training, 15% for validation, 15% for the testing respectively.
9. After training and determining the transition values. If the prediction equal to the original target class. Count it as matched value (xx), otherwise, count it as unmatched value (yy) which represent the stego target class.
10. Calculate the accuracy based on the matched and unmatched values (xx,yy) as follows.

$$Accuracy = (matched/(matched + unmatched)) * 100. \qquad (1)$$

11. Display the results of training model in NPR tool plots as they show the Performance, Training state, Error histogram, Confusion matrix, and Receiver operating characteristics shown in Fig. 4.

4   Results (164 Benchmarked Images)

This experiment is conducted with the NNPR training performance shown in Fig. 3. The total of 164 Bench-marked images from USC-SIPI and CVG have been trained. The results of the main plots are presented in Figs. 4, 5, 6, 7, and 8. The experiment results are depicted with Cross-Entropy, classification accuracy percent and error percent. The Cross-entropy plot shows results of appropriate classification that minimise cross-entropy. The lower value is the best one and zero means that there is no error. For the classification accuracy percent, it signifies the accurate classification. In the present study, the achieved classification accuracy is 91.4634%. At the start of the cross-entropy training, the plot depicts a maximum error. During the training, the network is presented. The validation is used to compute the simplification of the network. The testing phase is providing the network performance during and after the training.

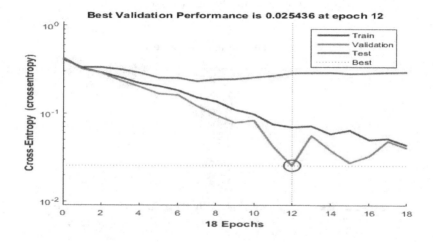

Fig. 4. Performance of CNN through (plotperform).

In Fig. 5. It plots the gradient of training state values and records during the training state.

In Fig. 6. The Error percent indicates the misclassification of samples fraction. When the error percent has a value equal to zero, it means there is no misclassification. If the error percent value is 100, it indicates the maximum of misclassification. In Fig. 6, most of the error values are close to zero with $+/-$ 0.04901 in a range of 20 bins between $+/-$ 0.9312.

In Fig. 7, the confusion matrices of training (114 images), testing (25 images), validation(25 images), and all confusion matrix show the number of classified and misclassified images in each class. The first and second classes represent the original and stego images respectively. The results give a significant classification of 82 images in each class.

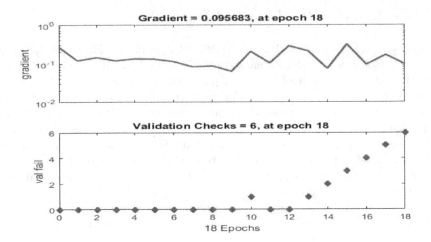

Fig. 5. Training State through (plottrainstate).

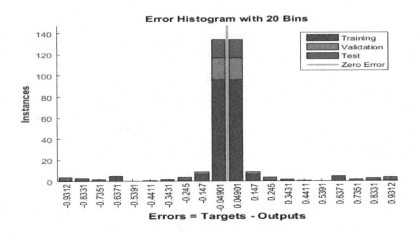

Fig. 6. Error Histogram through (ploterrhist).

Fig. 7. Confusion Matrix through (plotconfusion).

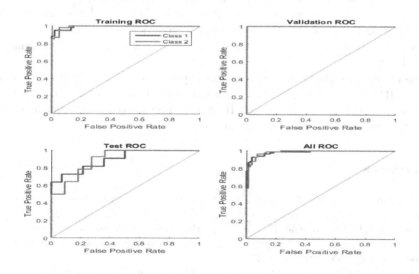

Fig. 8. Receiver Operating Characteristics through (plotroc).

In Fig. 8, the Receiver operating characteristics (ROC) is a plot that represents the quality of the classifier. It plots a curve to calculate the difference between (TPR) and (FPR). A bettter classification can be indicated by to what extent each curve hugs the left and top edge in the plot [6].

5   Conclusion

1. NNPR tool or Ensemble classifier proposed by Fridrich [20], are very efficient to classify some image steganography techniques like F5, Outguess mainly derived from spatial or frequency domain in images. However, image steganography in most of its current practice is robust enough. Meanwhile, this research aim only to study the detectability of image steganography through statistical machine learning measures and plots.

2. In addition, any highlight on the current challenges of this domain is quite important. This is achieved by sharing several experiments that contribute to image processing and analysis in both original and stego images. It discusses some of their applications. Correspondingly, for the detectability using classifier through the potential of NNPRtool experiment. However, this contribution followed many process of feature extraction including the study of different topics to find a suitable feature. Out of 34 local features and descriptors, some experiments involve matching 2 or more descriptors depending on its application. For instance, In objects and digits detection. Therefore, choosing significant descriptors are difficult and essential. In this research, it should takes into account that not all the descriptors such as (Harris (841), SURF (660), HOG(142884) vectors) are really useful on the detectability experiments with or without classifier because they result on a matrix of features and vectors not like lbp's operators which is computationally simple, and robust to analyse image in such challenging applications. Moreover, NNPR tool is a powerful tool for solving pattern recognition and classification problems as it help to create and train network and evaluate its performance efficiently.

3. One of the possible future work on this experiment is to study the potential of NPRTOOL to classify different images from other datasets with a better accuracy and training parameters. It emphasis the impact of machine learning in steganographic fields and experiments.

References

1. Cheddad, A., Condell, J., Curran, K., Mc Kevitt, P.: Digital image steganography: survey and analysis of current methods. Sig. Proc. **90**(3), 727–752 (2010)
2. Kadhim, I.J., Premaratne, P., Vial, P.J., Halloran, B.: Comprehensive survey of image steganography: techniques, evaluations, and trends in future research. Neurocomputing **335**, 299–326 (2019)
3. Lu, D., Weng, Q.: A survey of image classification methods and techniques for improving classification performance. Int. J. Remote Sens. **28**(5), 823–870 (2007)
4. Yadav, S.S., Jadhav, S.M.: Deep convolutional neural network based medical image classification for disease diagnosis. J. Big Data **6**(1), 1–18 (2019)
5. Deng, T.: A survey of convolutional neural networks for image classification: models and datasets. In: 2022 International Conference on Big Data, Information and Computer Network (BDICN), pp. 746–749. IEEE (2022)
6. Matlab-Team. https://www.mathworks.com/products/deep-learning.html

7. Matlab-Team (2023). https://www.mathworks.com/discovery/pattern-recognitio
   n.html
8. Wang, Y., Duncan, I.: Mitigating the threat of LSB steganography within data
   loss prevention systems
9. Guo, T., Dong, J., Li, H., Gao, Y.: Simple convolutional neural network on image
   classification. In: 2017 IEEE 2nd International Conference on Big Data Analysis
   (ICBDA), pp. 721–724. IEEE (2017)
10. Prabhu, V.S., Rajeswari, P., Blessy, Y.M.: A novel method for object recognition
    with a modified pulse coupled neural network. In: Sengodan, T., Murugappan, M.,
    Misra, S. (eds.) Advances in Electrical and Computer Technologies. LNEE, vol.
    711, pp. 521–531. Springer, Singapore (2021). https://doi.org/10.1007/978-981-
    15-9019-1_46
11. Jagtap, V.N., Mishra, S.K.: Fast efficient artificial neural network for handwritten
    digit recognition. Int. J. Comput. Sci. Inf. Technol. 5(2), 2302–2306 (2014)
12. Li, Q., Cai, W., Wang, X., Zhou, Y., Feng, D.D., Chen, M.: Medical image classi-
    fication with convolutional neural network. In: 2014 13th International Conference
    on Control Automation Robotics & Vision (ICARCV), pp. 844–848. IEEE (2014)
13. Sharma, N., Jain, V., Mishra, A.: An analysis of convolutional neural networks for
    image classification. Procedia Comput. Sci. 132, 377–384 (2018)
14. Chandra, M.A., Bedi, S.: Survey on SVM and their application in image classifi-
    cation. Int. J. Inf. Technol. 13, 1–11 (2021)
15. Li, Z., Yu, W., Cui, X.: Online classification of road roughness conditions with
    vehicle Unsprung mass acceleration by sliding time window. Shock Vibr. 2018
    (2018)
16. Stolte, S., Fang, R.: A survey on medical image analysis in diabetic retinopathy.
    Med. Image Anal. 64, 101742 (2020)
17. Bansal, D., Chhikara, R.: Performance evaluation of steganography tools using
    SVM and NPR tool. In: 2014 Fourth International Conference on Advanced Com-
    puting & Communication Technologies, pp. 483–487. IEEE (2014)
18. Mehta, P., Das, P., Bajpai, A., Bist, A.S.: Image classification using NPR and com-
    parison of classification results. In: 2016 3rd International Conference on Comput-
    ing for Sustainable Global Development (INDIACom), pp. 3231–3234. IEEE (2016)
19. Nissar, A., Mir, A.: Texture based steganalysis of grayscale images using neural
    network. Sig. Process. Res. 2(1), 17–24 (2013)
20. Friddrich, J.: https://dde.binghamton.edu/download/ensemble/ (2013)

Enhancing Image Quality of Aging Smartphones Using Multi-scale Selective Kernel Feature Fusion Network

Md. Yearat Hossain[✉], Md. Mahbub Hasan Rakib, Ifran Rahman Nijhum,
and Tanzilur Rahman

Department of Electrical and Computer Engineering, North South University, Dhaka,
Bangladesh
{yearat.hossain,mahbub.rakib,ifran.nijhum,
tanzilur.rahman}@northsouth.edu

Abstract. Smartphones are now the most popular medium for photography. Smartphones can capture better images than the hardware they use thanks to Machine Learning-based computational photography. Even though modern phones have sophisticated image enhancement applications, older and aged phones lag behind for a variety of reasons. Lens quality degradation, which results in washed out, soft-looking images, is one of the most common issues seen in older devices. We propose a method for improving such images by using an attention-based multi-scale residual neural network trained on a synthetic dataset. We chose two smartphones: an old device that captures degraded images and a modern flagship that provides reference enhanced images. Then we used the bloom filter, contrast, and highlight adjustment to make the reference images appear degraded. Later, we trained the model using synthetically degraded images and tested it on a variety of older devices. We achieved a maximum Peak Signal to Noise Ratio (PSNR) score of 74 throughout our experiments. To evaluate the model's images, we used Blind Image Quality Assessment (BIQA) methods such as HyperIQA. Aside from correcting the lens issue, the model achieves comparatively better sharpness, contrast, and color processing. Our proposed method generalizes well and achieves up to 11.45% improvement on novel devices by utilizing a very limited amount of data.

Keywords: Computational Photography · Image Enhancement · Image Processing

1 Introduction

Smartphones are now the most popular medium for photography. In 2023, approximately 54,400 images are captured every second [1]. To meet this ever-increasing demand, smartphone manufacturers are working hard to develop and launch more sophisticated camera systems on a regular basis. Despite numerous efforts to improve future camera systems, old devices fall behind for a variety of reasons. As a smartphone ages, the image quality it captures degrades due to a variety of issues. One of the most common issues that can degrade the photo quality of an aging smartphone is a damaged or blurry lens. It frequently results in the capture of blurry, gloomy, and washed-out images. Furthermore, sophisticated cutting-edge camera systems can render the previous generation of camera systems in smartphones obsolete.

Almost all modern smartphone manufacturers build their camera systems using computational photography. To capture and process images, computational photography heavily relies on computational algorithms rather than hardware systems. Because smartphones have such small sensors and lenses, this makes photography easier and more efficient. Machine Learning (ML) and Deep Learning (DL) algorithms, in addition to various computational image processing-related algorithms, are being developed and used in smartphones to improve overall image quality. Chip manufacturers are also incorporating dedicated neural processing units in their chips to support on-device machine learning applications [2].

Modern ML and DL algorithms have been found to perform significantly better on image processing and enhancement tasks [3, 4]. Despite the fact that much research has been conducted on image enhancement on smartphones and image enhancement in general, little research has been conducted on the problem we are addressing [5, 6]. Lens degradation can cause captured images to appear hazy and gloomy. Unwanted bloom effects could also be present in the captured images. Although some works deal with image dehazing and bloom issues, generalized solutions to multiple types of enhancement problems have yet to be introduced [7, 8]. Furthermore, companies tend to improve the camera systems on their latest devices while neglecting the older ones. However, by utilizing modern ML/DL techniques, we can also resolve several issues that are caused by old, outdated smartphones.

In our work, we used DL to solve the image degradation problem that occurs in aging smartphone devices due to lens degradation. Because lens quality degradation on old smartphones does not always result in the same type of image degradation, a model that can generalize image enhancement tasks is required. We used a multi-scale attention-based fully convolutional neural network to solve this problem. The model is well-known for its image enhancement capabilities, which we used to solve the problem we defined. Because the model transforms images, it requires two versions of the same image, one for the degraded image and one for the enhanced image. We used another smartphone with a good camera system to capture clean images, which we then processed to look similar to the original images captured by the degraded device. This assists us in producing image pairs of degraded and enhanced images. The model was then trained using these images, and it was tested using new images captured by the aged device. To achieve the best results, we combined the dataset with techniques such as patch generation and augmentation. We also used image blending to create the best-looking image possible from multiple outputs. On the model's outputs, we used no reference image quality assessment techniques, and the model always produced comparatively better images. Following the proposed method, the model significantly improves the image quality of the aged devices by using a small number of edited images. Along with producing clear, sharp images, the model learns to process color enhancement on images captured by the old device, effectively replicating a device with a much better camera system. Finally, we discussed the potential and future applications of our proposed method.

Our main contributions are as follows:

1. Proposing a generalized method for improving images that degrade on aging smartphones due to lens degradation.
2. Proposing a synthetic dataset generation method using a reference device for image enhancement problems.
3. Using the Blind Image Quality Assessment technique to evaluate image enhancement problems where no dataset exists.

2   Related Work

Image enhancement is a popular problem for researchers in the field of computer vision. Image enhancement techniques have improved dramatically over time thanks to ML and deep DL algorithms. Given a large dataset, the authors of [9] demonstrate that a multilayer perceptron can learn how to enhance and denoise dynamically from purposefully obscured and disturbed inputs. The authors of [10] proposed a Convolutional Neural Network (CNN) called MSR-Net to deal with the task of low light enhancement as a machine learning problem. Using machine learning techniques such as supervised learning and back propagation, the model could learn an end-to-end mapping between low light and enhanced images. In this study, the authors of [4] presented a unique hue-correction method based on the constant-hue plane in the RGB color space for the purpose of enhancing color images using deep learning. This paper [3] describes a novel technique called Zero-Reference Deep Curve Estimation (Zero-DCE). Their DCE-Net lightweight deep learning model predicts pixel-by-pixel and high-order curves for dynamic range correction of a given image for low-light image enhancement. Aside from these, Auto Encoder and Generative Adversarial Network (GAN) based research for low light image enhancement are also carried out [11, 12]. Deep Learning has grown in popularity in smartphone computational photography because it aids in overcoming the limitations imposed by the small hardware. Several papers that use DL for computational photography on smartphones have been published. The study [6] examines the first barrier to effective perceptual image enhancement, with a focus on the use of deep learning models on mobile devices. Their paper, which focuses on real-world photo enhancement, aimed to improve low-quality images taken with an iPhone 3GS smartphone to equivalent images taken with a DSLR camera. The authors of paper [5] explained the challenges of smartphone photography and how they compare to large DSLR cameras. They proposed using generative adversarial networks to map ordinary photos to DSLR-quality image quality. They used the Structural Similarity Index Measure (SSIM) and contextual losses as the content loss to maintain the realistic quality of an image. Finally, they used knowledge distillation to create a lightweight model that performs well in resource-constrained environments. Because of the increased demand, hardware manufacturers have also aided the field by developing better chips that allow smartphone devices to adapt to ML/DL-based applications. The authors of [2] conducted a thorough investigation into how neural processing units in various smartphone SOCs (System-on-a-Chip) performed in comparison to various desktop CPUs and GPUs. Their research described how the advancement of modern SOCs aided AI systems in the Android environment for on-device ML tasks. They created a benchmarking system that can provide scores for various SOCs performing ML tasks and thoroughly compared their results. They also demonstrated how modern SOCs are catching up to desktop grade systems on a variety of occasions, and how these advancements have aided in the development of stronger ML pipelines in smartphones.

3   Methodology

In the methodology section we discussed the model, dataset and the metrics used in our experiments. The workings of the model were discussed in the subsection titled "Model", while the generation of the dataset was covered in "Dataset". The evaluation metrics of the proposed system are discussed in the "Metrics" subsection.

3.1   Model

In this paper, we employ a model called MIRNet, which was first described in [13]. The model is a state-of-the-art Fully Convolutional Network (FCN) for various image enhancement tasks. The main advantage of this model is its ability to learn from both high and low-resolution versions of an image. The neural network is better able to learn spatially precise information at high resolutions, while contextual information is better learned at low resolutions. The model accomplishes this by adhering to a particular residual design paradigm, which dissects the images into different scales before relaying the data to subsequent layers. The MRB stands for "Multi-Scale Residual Block" and is the central part of the architecture. The primary responsibility of MRB is to use Selective Kernel Feature Fusion (SKFF) to learn information across parallel streams while preserving the original resolution throughout the network. Before reaching the SKFF modules, all MRBs contain Dual Attention Unit (DAU) blocks. To extract features from the convolution streams, the DAU suppresses less important information and shares information across streams while the SKFF module aggregates multi-scale branches. The architecture also introduces Recursive Residual Groups (RRG) where each RRG can contain multiple MRBs. For an image $I \in R^{H \times W \times 3}$ the extracted low-level features $X_0 \in R^{H \times W \times C}$ are achieved with the help of the convolution layer. These features are then passed to multiple numbers of RRGs to achieve the deep features $X_d \in R^{H \times W \times C}$. Then another convolution layer is applied on the X_d to obtain the residual image $R \in R^{H \times W \times C}$. Finally, the restored image $\hat{I} = I + R$ is obtained. The model uses Charbonnier loss [14] during the training phase.

3.2   Dataset

Each piece of data used in the model must be accompanied by two images: a degraded image and an enhanced target image. The model is trained to recognize the differences between the original and improved versions of an image down to the pixel level. We introduced a state-of-the-art high-end device capable of capturing high-quality images, as it is technically impossible to capture the same image in two different conditions using

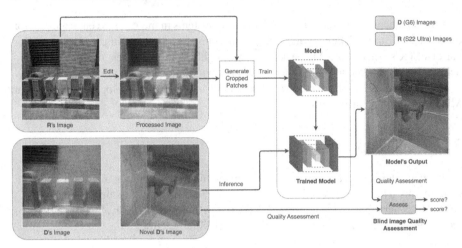

Fig. 1. The overall architecture of the proposed method. We captured similar images on both of the devices and then processed the **R**'s image to look degraded like the **D**'s image. We trained the model on cropped patches of the synthetic dataset and finally tested on novel images of **D**.

the same device. We used the notation **D** for the older device with the compromised camera and **R** for the device that captured the high-quality reference images used in the experiments. We took multiple shots of the same objects and scenes using both devices, trying to keep the framing consistent so that everything looked as close as possible from a spatial perspective. Then, we processed every image taken with the **R** by hand so that it looked as if it had been taken with device **D**. For this, we used Gimp [15], a widely-used, free, and open-source photo editor. Images captured by the faulty device could be artificially made to look like they came from a similar camera using in-built effects like a bloom filter, contrast, and highlight adjustment. The bloom filter generates soften looking images as the lights gently bleed into surrounding pixels and also makes the images slightly blurry. White noise filters wash out images while adjusting contrast and highlights help process them as precisely as possible. As a result, we managed to generate pairs of multiple images where the processed images acted as the degraded images and the original images captured by the **R** acted as the enhanced images. Later, during training, we cropped the high-resolution images to create numerous low-resolution image patches. Figure 1 depicts the overall architecture of the proposed method.

3.3 Metrics

During model training, we used Peak Signal to Noise Ratio (PSNR) (Eq. 2), which is derived from Mean Squared Error (MSE) (Eq. 1) as the performance metric. PSNR calculates the quality difference between the model's output and the corresponding input's ground truth. We had to evaluate the model's performance on novel images using blind no reference image quality assessment techniques because we didn't have any ground truth images for the actual degraded images captured by the aged devices. We used the method described in work [16], which calculates the sharpness of a given image based on the grayscale channel edges. Their proposed method can estimate image sharpness with no reference. We used HyperIQA [17] for overall image quality assessment, which is a deep learning-based method for Blind Image Quality Assessment (BIQA) that achieves state-of-the-art results in a variety of real-world image quality assessment datasets. Their method evaluates an image in three stages and employs a self-adaptive approach. To evaluate our images, we used a HyperIQA model that had been pre-trained on the KonIQ-10K IQA database [18]. The model is well generalized because it can evaluate the quality of various types of real-world images blindly. We used the term 'Quality' for the HyperIQA score and 'Sharpness' for the sharpness score obtained using the previously discussed method throughout our experiments. The HyperIQA method provides a score between 1 and 100, whereas the other method has no upper limit. In both cases, the higher the score, the superior the image to the other.

$$MSE = \frac{\sum_{M,N} [I_1(m,n) - I_2(m,n)]^2}{M \times N} \tag{1}$$

$$PSNR = 10\log_{10}(\frac{R^2}{MSE}) \tag{2}$$

4 Experimental Setup

We primarily used two different Android devices that were released five years apart. An old LG G6 acted as the D_1 and a Samsung S22 Ultra acted as the R to provide us with high quality reference images. The Sony IMX 258 sensor in the G6 device is 1/3.1 inch in size and shoots 13 MP images. The S22 Ultra features Samsung's flagship ISOCELL HM3 sensor, which shoots 108 MP images and measures 1/1.33 inch. The ISOCELL

Table 1. The specifications of the Android devices that we used in our experiments. When compared to the reference current generation high end device, the devices are 6, 8, and 9 years old, respectively. The Age indicates how long the device has been in use.

Brand	Model	Year	Sensor	OS	SoC	Age	Alias
Samsung	S22 Ultra	2022	ISOCELL HM3 108 MP, f/1.8, 23 mm, 1/1.33", 0.8 μm	12	Snapdragon 8 Gen 1 (4 nm)	1 year	R
LG	G6	2017	Sony IMX 258 13 MP, f/1.8, 30 mm, 1/3.1", 1.12 μm	9	Snapdragon 821 (14 nm)	6 years	D_1
Sony	Xperia Z2	2014	Sony Exmor RS 20.7 MP, f/2.0, 1/2.3", 1.12 μm	6.0	Snapdragon 801 (28 nm)	9 years	D_2
Samsung	J7	2015	13 MP, f/1.9, 28 mm	7.1	Snapdragon 615 (28 nm)	8 years	D_T

sensor is much larger and produces sharper images than the Sony IMX sensor. Then, to test the generalizability of our proposed method, we repeated the experiment with a Sony Xperia Z2 device that was 9 years old ($\mathbf{D_2}$). The Z2 device captures images that are slightly better than the G6 that we used in our experiment. Finally, we tested one of our trained models on images captured by a completely different device, a Samsung J7 ($\mathbf{D_T}$) that had never been used in any of our experiments. Table 1 lists all of the devices we used in our experiments. We used a desktop computer with a 3.70 GHz Intel i7 8700K CPU, 32 GB of DDR4 memory, and a 6 GB Nvidia RTX 2060 GPU to process the dataset and train the model.

5   Experiments and Results

First, we only took 10 images with our two devices, the $\mathbf{D_1}$ and $\mathbf{R}$. The photographs were taken in well-lit natural light conditions. We attempted to capture similar framed images in order to visually understand the differences between the images captured by each device. Then, using the $\mathbf{D_1}$ images as a guide, we applied some processing to the $\mathbf{R}$'s clean images. We attempted to replicate the overall look of the $\mathbf{D_1}$ images by using a combination of the bloom filter and a slight contrast-highlight adjustment. When the processing was completed, we were left with 10 edited images that appeared to be degraded versions of the 10 original clean images. The images were originally 2992×2992 pixels in size, and we cropped patches of 374×374 pixels from each image, allowing us to generate 64 images from each large image (Eq. 3). The dataset contained 640 pairs of images for the training process after the image patch generation was completed. We then cropped 128×128 patches at random because it worked best for our training setup. We had to use 3 RRG blocks, each of which contained 2 MRB blocks, for similar reasons. The model was then trained for 50 epochs using the Charbonnier Loss and Adam Optimizer, as described in the original paper.

$$PatchesPerImage = \frac{Original^H \times Original^W}{Patch^H \times Patch^W} \tag{3}$$

As the training process concluded, the model achieved a maximum PSNR score of 70. On our experimental setup, training the model took about 240 min. Table 2 displays the results of all experiments. The model was then put through its paces by inferring the novel full-sized (not cropped patches) images captured by the G6 device. All of the novel full-sized images tested on the models were completely unique; none of them were even used during the synthetic dataset generation process. The images were also taken in natural, well-lit lighting conditions. Figure 2 shows the model's output on novel captured images by the aged device. The model removes as much of the original image's glowing or bloom effect as possible while also improving the image's overall sharpness. Additionally, better contrast and color adjustment can be seen throughout the inferred images. The inferences were performed on the GPU, and the trained model took about 30 s to infer each full-sized image. Then we tried out different augmentations on the cropped patches. We used a similar process to generate cropped patches, but we added a 3-factor augmentation to each cropped patch. We generated $640 \times 3 = 1920$ image pairs by rotating each cropped image clockwise and anticlockwise. We trained the model

Fig. 2. Top: Novel images captured by the device D_1 (G6). Bottom: 1st Model's output on the same images.

using the same setup and obtained a similar PSNR score. However, the inferred results of this model are perceptually very different from the first. This model produces images that are slightly less sharp but have higher contrast and fewer artifacts. Nevertheless, the overall shadow enhancement capability declines (Fig. 3). As the number of datasets increases, the model improves at enhancing with fewer artifacts. Yet, our dataset lacked uniform balanced patches that represented all types of situations that needed to be fixed on a given image, such as shadow and highlight. As a result, the augmentation process steers the model toward producing slightly high contrast images. When compared to the second trained model, the first model produces better color and shadow enhancement but falls short on contrast and cleanliness. In this case, averaging the inferred images of the two models yields a significantly better result (Fig. 4). The averaged image achieves a good balance of the results produced by both models. To accomplish this, an image is passed through the two different models, and the inferred images are averaged using pixel-by-pixel operation to blend and generate a new image that is well balanced in every way. Finally, for the final experiment with D1, we added one more image to our main dataset to see how it affected the model. This time, the model was trained on 704 (11 × 64) cropped patches, and it achieved a PSNR score of 71. This model enhanced the color so that it appears more similar to the color processing of images captured by the device R (Fig. 5). However, the model performed poorly in darker regions because the newly added image did not contain much information that could aid in the shadow enhancement process. The performance of the model is directly related to how well the data is balanced to represent all of the information required by the model to learn.

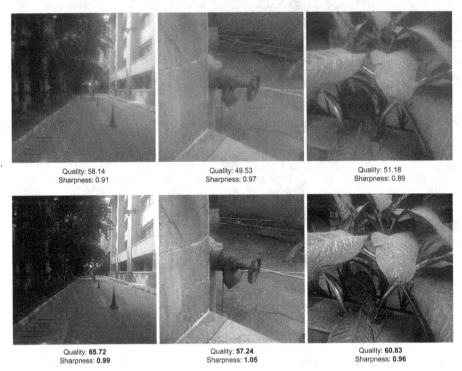

Quality: 58.14
Sharpness: 0.91

Quality: 49.53
Sharpness: 0.97

Quality: 51.18
Sharpness: 0.89

Quality: **65.72**
Sharpness: **0.99**

Quality: **57.24**
Sharpness: **1.05**

Quality: **60.83**
Sharpness: **0.96**

Fig. 3. Top: Novel images captured by the device D_1 (G6). Bottom: 2nd Model's output on the same images.

We also tested the generalizability of our proposed method on a 9-year-old Sony Xperia Z2 (D_2) with a similar type of lens issue. We captured 10 pairs of images, just like we did for the LG G6, and then edited the baseline R images to make them look degraded like the images captured by the D_2. The model achieves a PSNR score of 74 using the same procedure. Figure 6 depicts the end result of the image enhancement process. When compared to the images captured by the D_1, the D_2 images are slightly less degraded and thus achieve a comparatively better-looking enhancement. Finally, we tested our trained model on a completely new device, a Samsung J7 (D_T) released in 2015. Figure 7 depicts the inferred results. Given that it was trained on cropped images of only 10 large images, the model performs admirably. The model has never seen full-sized images and was trained on very small patches of synthetically processed images, but it produces very usable results. The model achieved sufficient generalization to handle novel images equally.

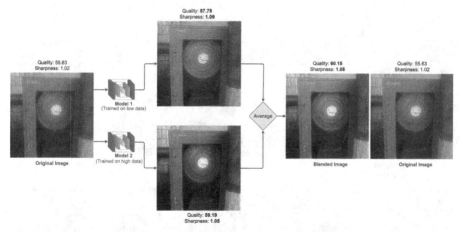

Fig. 4. An averaged blended image generates a much-balanced result compared to the individual models. Model that is trained on low data is good at sharpening but generates artifacts while the model trained on high amount of data generates better contrast.

Fig. 5. Gradually the model learns to enhance the color processing of given images replicating the ground truth images.

Fig. 6. Top row contains the original images captured by the D_2 (Xperia Z2). The bottom row contains the enhanced outputs of the model. The enhanced images look sharp and punchy compared to the originals.

Table 2. Results of all the experiments. The Enhancement Improvement percentage is based on the HyperIQA scores on novel test images.

Device	PSNR (Training)	Enhancement Improvement
D_1 (G6)	70	6.82%
D_1 (G6) with Augmentation	70	8.31%
D_2 (Z2)	74	7.70%
D_T (J7)	N/A	11.45%

6 Discussion

Even though we got a good result with only 10 images, the dataset generation can be challenging because the model's output is directly dependent on the training dataset. The dataset requires a proper balance of images with highlights, colors, and shadows. Cropping the images on lower dimensions would result in more data to train, but smaller patches could contain images that confuse the model, such as a plain white sky or a smooth monotonous surface. The experiments show that the model is not only attempting

Fig. 7. The top row represents the original images captured by the $\mathbf{D_T}$ (J7) and the bottom row displays the enhanced version of the corresponding image. Our proposed method generalizes well to work decently on completely novel devices too. Even though the model has never seen any samples of the J7, it enhances the images perfectly.

to correct lens-related issues, but also to improve contrast and color adjustment. The model is attempting to achieve a general enhancement procedure, but specific tasks would be much better performed by models that have been trained to solve specific problems. Furthermore, the image quality of the S22 Ultra is too good for the model to learn with such a small amount of data on the given setup. Images captured by the Xperia Z2 and Samsung J7 receive comparatively better enhancement because their original degraded images were already slightly better than images captured by the G6. We can see from the displayed samples in our work that the average HyperIQA score improvement for G6 is 7.56%, while it is 7.70% and 11.45% for Z2 and J7, respectively (Table 2). The more degraded the original images are, the harder it gets for the model to produce a better-looking image. We only intended to address a specific issue caused by faulty optics, but our proposed solution can be used to transfer color processing techniques from one device to another, allowing users to generate color processing based on their preferences.

We had to create a dataset from scratch because there was no pre-existing dataset for the specific problem. Though generating the synthetic dataset initially requires human expertise, we can estimate from our model's performance that the opposite operation can also be done using the model. For example, the model can be trained to learn how to degrade a good image, and the trained model can then be used to generate an increasing number of degraded images. This will aid in the generation of more synthetic degraded images in a semi-supervised manner, which can then be trained to build a much superior model with less effort.

As previously stated, due to the computational complexity, we were only able to experiment on a limited number of setups. The model requires a large amount of memory because it maintains the full-size resolution of the input image on multiple sections while performing other extended channel-based operations. However, training the model on higher resolution images would have allowed the model to learn more information from given patches. As discussed in the paper, increasing the RRG and MRB yields better results [13].

Even though the model requires a significant amount of computational power during the training process, inference of single images is relatively simple. Furthermore, by employing techniques such as model pruning, quantization, and knowledge distillation, a properly trained model can be significantly reduced in size. Smaller models can be deployed on modern phones, making them more future-proof. Older devices can also benefit from web-based off-device image processing applications. Because older devices lack the computational power of modern phones, web-based image processing can be built using our proposed method. Web-based image processing for older devices has the potential to unlock even more potential. Because the processing will take place on web servers, the phones will be able to receive long-term image processing support by updating models on various occasions.

7 Conclusion

In our paper, we proposed using a multi-scale attention-based feature fusion network to improve the image quality of aging smartphones. We captured images with two smartphones and generated synthetic degraded images that resembled the aged device. Then, we used small patches from the original large images to generate dataset pairs for training the model, and our proposed method performed well with only 10 edited images. We used no reference BIQA methods to evaluate the inferred outputs because there were no ground truth images to compare them to. The model not only learns to fix lens issues, but it also learns to process image color and contrast like a flagship smartphone. The future works could include the on-device performance of the trained model, the generation of semi-supervised datasets, and an ablation study of the proposed system using more computational resources.

References

1. How Many Photos Are There? 50+ Photos Statistics (2023). https://phututorial.com/photos-statistics/. Accessed 14 Jan 2023
2. Lee, J., et al.: On-device neural net inference with mobile GPUs. arXiv preprint arXiv:1907. 01989 (2019)
3. Guo, C., et al.: Zero-reference deep curve estimation for low-light image enhancement. In: IEEE/CVF Conference on Computer Vision and Pattern Recognition (CVPR), pp. 1780–1789 (2020)

4. Kinoshita, Y., Kiya, H.: Hue-correction scheme based on constant-hue plane for deep-learning-based color-image enhancement. IEEE Access **8**, 9540–9550 (2020)
5. Hui, Z., Wang, X., Deng, L., Gao, X.: Perception-preserving convolutional networks for image enhancement on smartphones. In: Leal-Taixé, L., Roth, S. (eds.) ECCV 2018. LNCS, vol. 11133, pp. 197–213. Springer, Cham (2019). https://doi.org/10.1007/978-3-030-11021-5_13
6. Ignatov, A., et al.: PIRM challenge on perceptual image enhancement on smartphones: Report. In: European Conference on Computer Vision (ECCV) Workshops (2018)
7. Engin, D., Genc, A., Kemal Ekenel, H.: Cycle-dehaze: enhanced CycleGAN for single image dehazing. In: IEEE Conference on Computer Vision and Pattern Recognition Workshops, pp. 825–833 (2018)
8. Kuanar, S., Mahapatra, D., Bilas, M., Rao, K.R.: Multi-path dilated convolution network for haze and glow removal in nighttime images. In: Visual Computer, pp. 1–14 (2022)
9. Burger, H.C., Schuler, C.J., Harmeling, S.: Image denoising: can plain neural networks compete with BM3D? In: IEEE Conference on Computer Vision and Pattern Recognition (CVPR), pp. 2392–2399 (2012)
10. Shen, L., Yue, Z., Feng, F., Chen, Q., Liu, S., Ma, J.: MSR-net: low-light image enhancement using deep convolutional network. arXiv preprint arXiv:1711.02488 (2017)
11. Lore, K.G., Akintayo, A., Sarkar, S.: LLNet: a deep autoencoder approach to natural low-light image enhancement. Pattern Recogn. **61**, 650–662 (2017)
12. Yan, L., Fu, J., Wang, C., Ye, Z., Chen, H., Ling, H.: Enhanced network optimized generative adversarial network for image enhancement. Multimedia Tools Appl. **80**, 14363–14381 (2021)
13. Zamir, S.W., et al.: Learning enriched features for real image restoration and enhancement. In: Vedaldi, A., Bischof, H., Brox, T., Frahm, J.-M. (eds.) ECCV 2020. LNCS, vol. 12370, pp. 492–511. Springer, Cham (2020). https://doi.org/10.1007/978-3-030-58595-2_30
14. Charbonnier, P., Blanc-Féraud, L., Aubert, G., Barlaud, M.: Two deterministic half-quadratic regularization algorithms for computed imaging. In: 1st International Conference on Image Processing, vol. 2, pp. 168–172 (1994)
15. GIMP - GNU Image Manipulation Program. https://www.gimp.org/. Accessed 24 Feb 2023
16. Kumar, J., Chen, F., Doermann, D.: Sharpness estimation for document and scene images. In: 21st International Conference on Pattern Recognition (ICPR 2012), pp. 3292–3295 (2012)
17. Su, S., et al.: Blindly assess image quality in the wild guided by a self-adaptive hyper network. In: IEEE/CVF Conference on Computer Vision and Pattern Recognition, pp. 3667–3676 (2020)
18. Hosu, V., Lin, H., Sziranyi, T., Saupe, D.: KonIQ-10k: an ecologically valid database for deep learning of blind image quality assessment. IEEE Trans. Image Process. **29**, 4041–4056 (2020)

Off-Line Handwritten Math Formula Layout Analysis by ConvGNN

Kawther Khazri Ayeb and Afef Kacem Echi[(✉)][iD]

National Superior School of Engineering, University of Tunis, LR: LaTICE, Tunis,
Tunisia
ff.kacem@gmail.com

Abstract. This work aims to construct an end-to-end system that can accurately analyze the layout of a mathematical formula. To accomplish this, we proposed a deep-learning architecture that makes use of a Convolutional Graph Neural Network (ConvGNN)—a variant of a convolutional neural network that operates directly on graphs to label symbols and their spatial relationships within the context of mathematical formulas, while also factoring in the features of the neighbors and their interrelationships. Testing different models of ConvGNN and comparing with the classical method: a Multi Layers Perceptron (MLP), we confirm the effectiveness of ConvGNN, especially the spectral-based one, as a suitable way to model formula structure that a graph can represent. We demonstrate that our system competes with some related works in several experiments on CROHME, a handwritten math formula standard database. These tests revealed an accuracy of 82% for MLP, 23% for simple ConvGNN, and 92% for spectral-based ConvGNN.

Keywords: Math Formula Recognition · Deep Learning · Structure analysis · ConvGNN

1 Introduction

There needs to be more digital access to many scientific and engineering texts that contain essential mathematical knowledge. Other newer materials are scanned journals and papers, word proceeds and type-set files, PDF files, and Web pages, which are often difficult to get sources. Mathematical information in these documents is hard to use or find because it is not easy to search for or get to. One will need an equation editor manually retype the formulas into a computer's algebra system or another program that accepts mathematical input to use this data. It is becoming increasingly important to find ways of extracting and communicating mathematics that facilitates automatic processing, searching, indexing, and reusing in other mathematical applications and contexts. Keep in mind the many ways in which mathematics differs from regular writing. The two-dimensional structure (fraction bars, roots, large delimiters, etc.), the frequent font changes, the various symbols (alphabetic characters, Greek letters, numerals, math operators, etc.), and the widely varying notational conventions among different sources are all factors. Handwritten mathematics presents a complex recognizing problem due to noise and merged or broken characters.

Symbol recognition and structural analysis are the two main steps of the math formula recognition issue. In the first step, symbols from the input formula are extracted. It returns a collection of symbols and information about them (position, size, etc.). A classifier is then used to establish the identity attribute of the symbol. The difficulty in adequately recognizing the symbol could lead to confusion and mistakes. The next step is examining the formula layout, which can be very challenging in mathematics because it often requires much work to determine the precise relationship between two or more symbols. Therefore, even when all the constituent symbols are recognized correctly, recognizing a math formula is more complex.

As emphasized in [2] and [3], few papers addressed specific math formula recognition problems. Additional researchers have begun to focus on this topic very recently. Much work has been put into analyzing and recognizing mathematical formulas, but it still needs to be completed. Classical recognition methods have succeeded but generally impose assumptions or restrictions on the problem to make it manageable. Several areas still need work, including ambiguity resolution, contextual information utilization, error correction, and performance assessment. Moreover, more functional issues would surface as we include such math formula recognition systems in practical applications. Typical machine learning methods process this type of data into a vector of real values, which causes information to be lost. Researchers have made more significant efforts to take advantage of deep learning techniques for enhancing the performance of math formula recognition systems in light of recent successes of deep learning in various applications, including natural language processing and image processing.

This paper presents an end-to-end automatic system for reading, parsing, and reusing handwritten mathematical formulas. We introduced a Convolutional Graph Neural Network (ConvGNN) as a deep learning architecture for determining mathematical symbols' lexical units and spatial arrangement. To deal with the multi-relational data typical in practical knowledge bases, [1] has created ConvGNNs, similar to a recent family of neural networks operating on graphs. We shall show that ConvGNN is a good model for analyzing the structure of the mathematical formula, as graphs have a greater expressive power and formulas can be understood as a collection of symbols and relationships between them (see Fig. 1). Our study of this framework's application to this problem is the first of its kind.

The remaining sections of the paper are as follows. In Sect. 2, we look at some concurrent works. Afterward, we focus on recognizing math formulas that must be overcome during symbol recognition and the analysis of formula structure in Sect. 3. We close the paper with experiments in Sect. 4, concluding remarks, and some prospects in Sect. 5.

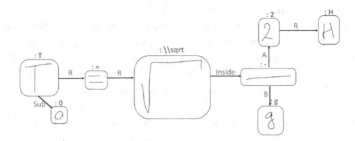

Fig. 1. A formula represented by a graph: Symbols serve as nodes, relations denoted by their types as edges, and entity classes as labels for the nodes. (e.g., letter, RS, HFB). The red labels on the edges and nodes indicate the information that must be inferred. (Color figure online)

2   The State of the Art

[6] has compiled a summary of feature extraction methods and classifiers for classification for off-line math symbol and formula recognition (see Table 1).

Table 1. Summary of feature extraction methods and classifiers for math symbols and formula recognition.

Refs.	Features set	Classifier	Reco. Rate
[7]	Zonal, Structural, Skeleton based and Directional features	SVM	85%
		MLP	92%
[8]	Projection histogram	SVM	97.58% and 98.40% for two different datasets
[9]	Normalized chain code, Moment invariant features, Density feature, Projection histogram	SVM	93.8%
[10]	Centroid and bounding box	ANN	–
[11]	PHOG	SVM	96%
[12]	Character Geometry	SVM	88%
		KNN	91%

Regarding this study, it was found that the Supported Vector Machine (SVM) is a practical classifier and that the projection histogram and the Pyramid of Histogram Oriented Gradients (PHOG) are the most pertinent features that provide improved accuracy. But, the used feature extraction methods are symbol dependent. Depending on the variety of symbols used within math formulas, these methods must be verified for the large set of mathematical symbols. Many classifiers are tested over the features extracted from several feature extraction methods. So the result of classification and recognition depends on the type of features used for training the symbols within math formulas. In addition, most of the classification methods are based on a particular category of math formulas. These methods must be tested over various math formulas of different types like linear equations, algebraic equations, geometric equations, etc. Public datasets of math formulas have been unavailable for the past few years. Thus, researchers compiled their dataset, which may confine the issue to a particular domain.

All the above systems require extracting the features before the classification step. Many methods, such as Convolutional Neural Networks (CNNs) and Recurrent Neural Networks (RNNs), no longer have this limitation due to the recent advancements of deep learning in computer vision. They can eliminate the need for human feature extraction and selection through their ability to extract features automatically, learn, and classify them. Instead of preprocessing the data to get features like textures and forms, a CNN takes the raw pixel data as input, learns how to extract these features, and then figures out what object they belong to. However, many types of real-world data are represented as graphs or networks. That is especially true of data describing complex structures like social networks, communication networks, knowledge graphs, road maps, molecules, etc. Thus, standard deep learning architectures like CNNs and RNNs can not be used to model such data types. This motivation has led to the development, in recent years, of several techniques that, for graph data, reconceive the idea of convolution. These developments have been boosted by the successes of CNN in practical applications.

In his review, [4] traced the developments in applying neural networks to directed acyclic graphs, which prompted the earliest research on Graphical Neural Networks. (GNNs). These preliminary investigations are classified as recurrent graph neural networks (RecGNNs). The representation of a target node is learned by propagating neighbor information iteratively until a stable fixed point is reached. According to [4], this process is computationally expensive, and there have been increasing efforts in recent years to surmount these obstacles. Recent findings regarding GNNs have led to ConvGNNs, also named Graphical Convolutional Networks (GCNs), an excellent domain for studying problems common to many documents and graphics recognition. We opted for such deep learning networks to recognize math formulas especially. In most existing math formula recognition methods, symbol segmentation and recognition and formula structure analysis are solved sequentially to provide a mathematical interpretation. However, the primary disadvantage of this sequential method is that symbol segmentation and classification errors will propagate to structural analysis. Symbol recognition and structural analysis are assumed to be separate tasks. However, these three tasks are intrinsically interdependent-the global structure aids in recognizing symbols by humans and vice versa. We proposed using contextual information when recognizing math symbols and their spatial arrangement to improve the math formula recognition system performances. As modeling and using such information is still challenging and the standard deep learning architecture like CNNs and RNNs do not work, we are oriented to ConvGNN, as it will be explained next.

3   Proposed System

In Fig. 2, we display the ConvGNN's architecture mainly proposed for formula structure analysis, where the network's input is the entire formula's graph. The nodes represent the symbols, while the links represent their spatial relationships. To ensure link prediction, we also proposed adding nodes for links. The output of the proposed ConvGNN is a labeled graph where symbols are classified into lexical units, and their relationships are identified.

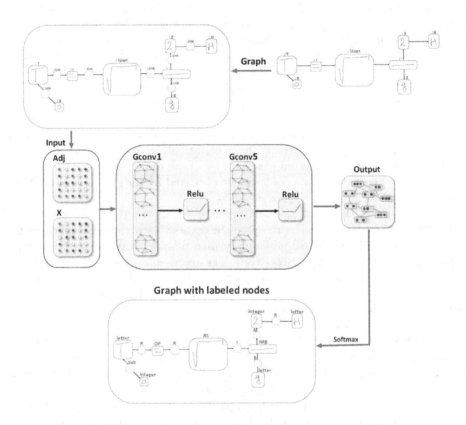

Fig. 2. ConvGNN's architecture for math symbol classification and formula structure analysis.

Note that an image is a particular case of a graph in which pixels are linked by pixels next to them. In 2D convolution, each pixel in an image is like a node in a graph. The size of the filter determines neighbors. The weighted average pixel values of the red node and its neighbors are used in the 2D convolution. A node's neighbors are ordered and have a defined size (see Fig. 3). Similar to the 2D convolution, the graph convolution utilizes the weighted average of a node's surrounding information. One chooses the average value of the red node's and its neighbors' node features to obtain a hidden representation of the red node. In contrast to image data, a node's neighbors are unordered and of variable size (see Fig. 3).

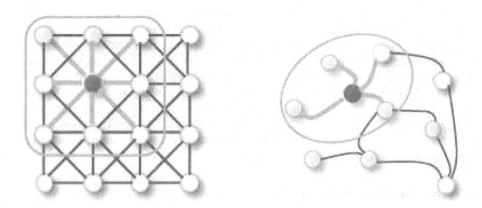

Fig. 3. 2D Convolution vs. Graph Convolution [4].

In what follows, we give a brief definition of the ConvGNN. Then, we explain how to propagate information through its hidden layers. We explain how the ConvGNN accumulates information from preceding layers and provides usable graph node feature representations.

3.1   ConvGNN Definition

ConvGNNs process graphs, and it extends grid convolution to graph data. Aggregating a node's features and neighbors' features yields high-level node representations. Given a graph $G = (V, E)$ or $G = (A, X)$, it inputs:

- a feature matrix X of size $N \times F^0$, where N is the number of nodes and F^0 is the number of input features for each node.
- an adjacency matrix A of size $N \times N$, representing the graph structure.

A hidden layer in the ConvGCN can thus be written as

$$H^i = f(H^{i-1}, A) \tag{1}$$

where $H^0 = X$ and f is a propagation rule. Each layer H^i matches to an $N \times F^i$ feature matrix where each row is a feature representation of a node. The propagation rule f aggregates these features into the following layer's features. Each layer abstracts features. ConvGNN variants differ solely in propagation rule f. We started with a simple propagation rule-based convGNN, and spectral rule testing followed.

3.2 Simple ConvGNN

One of the simplest possible propagation rule is

$$f(H^i, A) = \sigma(AH^iW^i) \tag{2}$$

here W^i is the weight matrix for layer i, and σ is a non-linear activation function (the ReLU function). The weight matrix has dimensions $F^i \times F^{i-1}$. In other words, the number of features at the next layer is determined by the size of the second dimension of the weight matrix. Note that the simple propagation rule represents a node as the sum of the feature representations of its neighbors. That has two major drawbacks: the aggregated representation of a node does not include its features, and nodes with high degrees will have high values in their feature representation. On the other hand, nodes with small degrees will have small values, which can cause problems with growing gradients and make it harder to train with algorithms like stochastic gradient descent, which are sensitive to feature scaling.

3.3 Spectral-Based ConvGNN

We explain why a specific graph-based neural network model $f(X, A)$ that we will use in the rest of this paper makes sense from a theoretical point of view. We look at a ConvGNN with the following rule for layer-by-layer propagation:

$$f(X, A) = \sigma(D^{-0.5}\hat{A}D^{-0.5}XW) \tag{3}$$

- D is the degree matrix: A diagonal matrix with information about each node's degree (the number of edges that connect to it). It is used with the adjacency matrix to build a graph's Laplacian matrix.
- $\hat{A} = A + I$ is the adjacency matrix with added self-connections (I: identity matrix) to take into account the feature of the node itself and not only its neighbor's features and to avoid this limitation encountered in the simple ConvGNN.
- W: the weight matrix
- σ : the ReLU activation function.

The only thing that makes the spectral rule different from the simple rule propagation is the choice of the aggregate function. Using the simple rule, here's how to figure out the aggregate feature representation of the ith node:

$$aggregate(A, X)_i = A_iX = \sum_{j=1}^{N} A_{i,j}X_j \tag{4}$$

As shown in the equation, the contribution of each neighbor is wholly based on the neighborhood that the adjacency matrix A defines. In the case of the spectral rule, the aggregate feature representation is computed as follows:

$$agregate(A, X)_i = D^{-0.5}A_iD^{-0.5}X = \sum_{j=1}^{N} D_{i,i}^{-0.5}A_{i,j}D_{j,j}^{-0.5}X_j \tag{5}$$

This propagation rule's form can be explained using a first-order approximation of localized spectral filters on graphs. As done by [1], we first used a Chebychev polynomial approximation. We then proposed using other approximation polynomials and found that the Hermite polynomial gives better results. We look at spectral convolutions on graphs, which are defined as the multiplication of a signal $x \in \mathbf{R}^N$ (a scalar for each node) by a filter $g_\theta = diag(\theta)$ with a parameter $\theta \in R^N$ in the Fourier domain:

$$g_\theta \times x = U_{g_\theta} U^T x \tag{6}$$

where U is the matrix of eigenvectors of the normalized graph Laplacian $L = I_N - D^{-1/2} A D^{-1/2} = U \wedge U^T$, with a diagonal matrix of its eigenvalues $\wedge$ and $U^T x$ being the graph Fourier transform of x. Thus, g_θ is a function of the eigenvalues of L, i.e., $g_\theta(\wedge)$. Evaluating the preceding equation is computationally costly because multiplication with the eigenvector matrix U is $O(N^2)$ expensive. Furthermore, computing the eigendecomposition of L in the first place might be prohibitively expensive for large L graphs. To face this problem, the Chebyshev polynomials are recursively defined as:

$$T_k(x) = 2x T_{k-1}(x) - T_{k-2}(x) \tag{7}$$

With $T_0(x) = 1$ and $T_1(x) = 2x$. The Hermite polynomials are recursively defined as

$$T_k(x) = x T_{k-1}(x) - (k-1) T_{k-2}(x) \tag{8}$$

With $T_0(x) = 1$ and $T_1(x) = x$.

4   Experiments

4.1   Dataset and Metrics

We tested our model on CROHME 2019 [5], a public dataset for handwritten math formulas. The math formulas are marked at the symbol level by giving each symbol node a ground truth label from a list of 20 labels (see Table 2 for more information about the labels).

For the experiments, we split the used math formula dataset, composed of 180 formulas into a training, validation, and testing set, using 50% (about 90 math formulas generating 1737 train nodes), 20% (about 36 math formulas generating 669 validation nodes) and 30% (about 54 math formulas generating 1096 test nodes), respectively. We chose to keep only 50% for training and an additional 30% for testing because we want to see how well our system can adapt even with a small training set. We used the accuracy metric, the number of correctly predicted sequences, to judge how well our math formula recognition system worked.

Table 2. Used labels.

Label	Signification
letter	Letters from a (resp. A) to z (resp. Z)
P	Operator (+, −, <, >, /, etc.)
DL	Delimiter ((,), , , etc.)
Integer	digits from 0 to 9
HFB	Horizontal Fraction Bar
IS	Integral Sign
MS	Math Symbol ($\triangle$, ∞, $\emptyset$, etc.)
RS	Root Symbol
Pun	Punctuation sign (,, ..., etc.)
SPS	Summation Product Symbol
NF	Name of Function (f(x), g(x), etc.)
MF	Math Function (cos, sin, etc.)
LS	Limit Sign
Det	Determinent
R	Right
A	Above
B	Below
I	Inside
Sub	Subscript
Sup	Superscript

4.2   Experimental Setup

The proposed ConvGNN is a 5-layer network. We trained it on the architecture described in Sect. 3 in a semi-supervised way. We evaluated it on a test set of 54 math formulas from which we generated 1096 symbol and relation nodes to classify.

We used the Hermite filters of order 2 ($k = 2$), and the initial number of units per hidden layer is 18. The number of classes is 20. We used the $L2$ regularization factor of 5.10^{-4} for the ConvGNN layer and a dropout of 0.005. We trained the model as a single big graph batch for a maximum of 1000 epochs with a learning rate of 0.001 and early stopping with a window size of 50. That means we stopped training if the validation loss didn't go down for 50 epochs in a row.

4.3   Obtained Results

Test Accuracy: We run our tests on four different models: Multi-layer Perceptron (MLP) model, the simple ConvGNN model, and the spectral one using, respectively, Tchebychev (ConvGNN-Hermite) and Hermite (ConvGNN-Chebyshev) polynomial approximation filter. The results of our studies to compare and evaluate are shown in Table 3. Our classification accuracy on test nodes, run time, and cost are reported.

Our results successfully demonstrate the excellent performance of spectral ConvGNN based on the Hermite approximation filter, even though we used limited nodes extracted from a part of the CROHME dataset. In addition, Spectral

Table 3. Results from simple and spectral ConvGNNs.

Proposed Models	Accuracy	Runtime	Cost
MLP	0.82	0.00698	0.62923
Simple ConvGNN	0.28	0.03124	2.17124
ConvGNN-Hermite	0.92	0.02992	0.31295
ConvGNN-Chebyshev	0.82	0.02992	0.68176

ConGNN can perform better in minor training epochs. Our model is more sophisticated than MLP, and each training epoch has many parameters to learn. Hence our training time is longer.

Accuracy, Loss Change with Training Epoch: Figure 4 and Fig. 5 demonstrate the training and validation accuracies and losses in each training epoch.

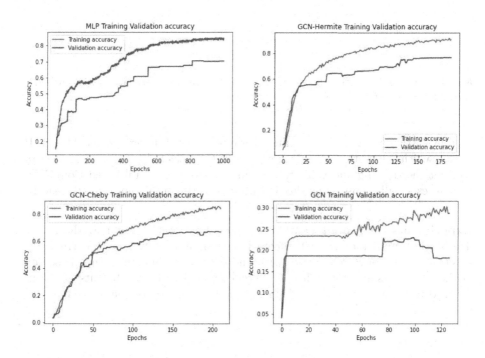

Fig. 4. Training Validation Accuracy.

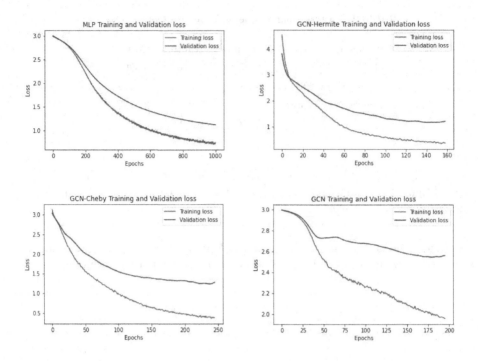

Fig. 5. Training Validation Loss: (a) MLP, (b) ConvGNN-Hermite, (c) ConvGNN-Cheby (d) Simple ConvGNN.

Our system can get good results after only a few training epochs. The MLP model, which is less stable, needs more than 600 epochs to be stable. Also, we can see that the training and validation accuracy/loss of Spectral-based ConvGNN models, especially those using the Hermite Filter, rise or fall quickly and steadily.

Experiments on Model Depth. In these tests, we look at how the model depth (the number of layers) affects how well it can classify. We tried our spectral-based ConvGNN model with the same experimental setup presented in the previous subsection with a different number of layers. Results are summarized in Fig. 6. A 5-layer model gives the best results for the dataset in this work.

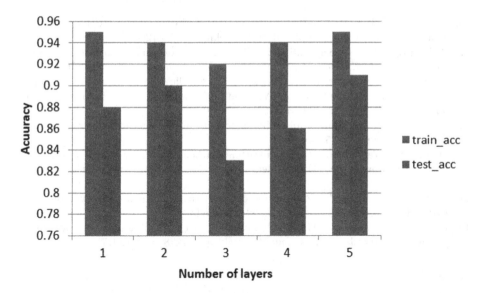

Fig. 6. Impact of model depth on classification accuracy.

5 Conclusion and Future Work

This paper introduced a novel math formula layout analysis approach: a Spectral-based ConvGNN. Experiments on several math formulas extracted from CROHME datasets prove that the proposed ConvGNN model can encode both graph structure and node features helpfully for math symbol classification and spatial relation identification. Preliminary tests on our method are promising since the results outperform several proposed methods based on classical approaches while being computationally efficient. The presented approach has several potential applications, including further research into representation learning for graphs and its use in recognizing mathematical formulas.

References

1. Kipf, T.N., Welling, M.: Semi-supervised classification with graph convolutional networks. In: ICLR (2017)
2. Dimitriadis, Y.A., Coronado, J.L.: Towards an ART based mathematical editor, that uses online handwritten symbol recognition. Pattern Recogn. **28**(6), 807–822 (1995)
3. Lee, M.C.: Understanding mathematical expressions using procedure-oriented transformation. Pattern Recogn. **27**(3), 447–457 (1994)
4. Zonghan, W., Shirui, P., Zhang, C.: A comprehensive survey on graph neural networks. arXiv:1901.00596v4. Accessed 4 Dec 2019
5. Mouchère, H.: ICDAR 2019 competition on recognition of handwritten mathematical expressions and typeset formula detection. In: International Conference on Document Analysis and Recognition (ICDAR) (2019)

6. Vaz, T., Vernekar, N.: A survey on evaluating handwritten iterative mathematical expressions. Int. J. Res. Appl. Sci. Eng. Technol. **6**(IV) (2018)
7. Padmapriya, R., Karpagavalli, S.: Offline handwritten mathematical expression recognition. Int. J. Innov. Res. Comput. Commun. Eng. **4**(1), 52–59 (2016)
8. Gharde, S.S., Baviskar, P.V., Adhiya, K.P.: Identification of handwritten simple mathematical equation based on SVM and projection histogram. Int. J. Soft Comput. Eng. **3**(2), 425–429 (2013). ISSN 2231-2307
9. Bharambe, M.: Recognition of offline handwritten mathematical expressions. Int. J. Comput. Appl. **2015**, 35–39 (2015)
10. Shinde, S., Waghulade, R.: Handwritten mathematical expressions recognition using back propagation artificial neural network. Commun. Appl. Electron. **4**(7), 1–6 (2016)
11. Jimenez, N., Nguyen, L.: Recognition of handwritten mathematical symbols with PHOG features (2013). http://cs229.stanford.edu
12. Fathima, I., Ashoka, K.: Machine learning approach for recognition of mathematical symbols. Int. J. Sci. Res. Eng. Technol. **6**(8), 2278–0882 (2017)

A Semi-supervised Teacher-Student Model Based on MMAN for Brain Tissue Segmentation

Salah Felouat[1] , Ramzi Bougandoura[1], Farrah Debbagh[1],
Mohammed Elmahdi Khennour[1(✉)], Mohammed Lamine Kherfi[1,2],
and Khadra Bouanane[1]

[1] Laboratoire d'intelligence Artificielle et des Technologies de l'information, Kasdi Merbah University Ouargla, Ouargla, Algeria
mahdi.kh5.com@gmail.com

[2] LAMIA Laboratory, Université du Québec à Trois-Riviéres, Trois-Riviéres, Canada

Abstract. In the medical field, deep learning technologies help radiologists gain accurate diagnoses in evaluating both neurological conditions and brain diseases. But these technologies require a huge amount of data to train and validate. The acquisition and labeling of these data are pretty expensive and require medical experts. Therefore, it is necessary to build models that do not require a lot of labeled data like semi-supervised methods. These methods train on few samples of data in a supervised manner and extend this knowledge to the rest of the unlabeled samples. For this reason, we propose a semi-supervised technique called Teacher-Student model using Multi-Modal Aggregation Network (MMAN) and apply it for brain tissue segmentation from Magnetic Resonance Images (MRI). The dataset (MRBrains Challenge) contains a small portion of labeled data and the majority of them are unlabeled. Our proposed method consists of two main steps. The first one aims to exploit the huge unlabeled data to increase the volume of the training set. The second step, it follows the strategy of the Teacher-Student technique in an iterative manner to enhances the performance of our model. The segmentation process will divide the brain into gray matter, white matter and cerebro-spinal fluid. Our approach allows for an improved prediction of brain image segmentation to reach a mean accuracy of 96.21%.

Keywords: Medical Image Segmentation · Semi-Supervised Learning · MMAN · Teacher-Student Model · Pseudo-Labeling

1 Introduction

Recently, medical development has witnessed a great progress specifically in computer vision. The emergence of deep networks has played an important role in developing a solution to several problems that the medical field was suffering from; including those related to medical image processing techniques, image

A. Bennour et al. (Eds.): ISPR 2023, CCIS 1940, pp. 65–75, 2024.
https://doi.org/10.1007/978-3-031-46335-8_6

classification and segmentation, which is not an easy task in fact. There are several problems that face this field, although architectural advances have reached high performance, they still need enormous and high-quality labeled data to train an image segmentation model. The acquisition of this data requires human experts to segments each slide of data manually. This process is pretty expensive nonetheless time consuming, especially for pixel-level annotation. One of the solutions that is used to handle this problem and minimize the cost of segmenting datasets is semi-supervised learning, which use a small set of labeled data and large set of unlabeled ones. SSL plays an important role in the medical field especially in medical image segmentation, where it helps radiologist in diagnosis of brain diseases such as Alzheimer's disease, dementia and epilepsy [1–3]. Magnetic Resonance Imaging (MRI) is one of the most common and interesting scans in the medical field which can bring an accurate scan of brain tissues. Despite the majority of proposed model uses single-modal of MRI scans, multi-modal's models harness these modalities to extract complementary information from the scans to increase the accuracy and performance of segmentation.

Several previous work in medical image segmentation such as the works of [4–7] rely on supervised learning approach. While for semi-supervised ones, they use images from a single scan(modal) [8]. In the work of [9], the author relied on naive-student approach to leverage the SSL model for video segmentation. The work of [10] focuses on semantic image segmentation based on weakly and semi-supervised learning. In the work of [11], they use a dilated CNN to produce dense object localization to enhance weakly and SSL segmentation.

In this paper, we propose a semi-supervised model based on Deep Convolutional Neural Networks which proved its effectiveness regarding the performance of segmenting major brain tissues such as Cerebro-Spinal Fluid (CSF), White Matter (WM) and Gray Matter (GM). Our work is based on one of the most successful models that have been designed for medical image segmentation during the MRBrains challenge [12] the Multi-Modal Aggregation Network (MMAN) [13]. This network is able to learn from multi-modality MRI an accurate, fast and multi-scale features with complementary information, which grant it the 2nd place in the MRBrains challenge[1]. The model that won the 1st place has better accuracy with cost of slow training, which will impact the performance of our model if we choose it.

Our contribution aims to use the MMAN to extract pseudo-masks of the unlabeled data set. Then, use them to train brain image segmentation models, in addition to applying the teacher-student method in an iterative context to train our final model for more accurate labeling of the unlabeled data in brain tissue segmentation from MRI. The proposed model has achieved a comparable results compared to the original work of [13], where the author used a supervised approach on the same dataset, which indicate the generalization capability of our approach while using less annotated data.

Throughout this paper, we will refer to segmentation mask as label; hence, pseudo-mask to pseudo-labeling. In the following, we start by giving an overview

---

[1] https://mrbrains13.isi.uu.nl/results/.

of the model and its structure in Sect. 2. In Sect. 3, we describe the conducted experiment, its setup and discuss the obtained results. Finally, we conclude our work in the 4th section.

2 Model Overview

Semi-supervised learning (SSL) is a machine learning approach based on the process of training a model using labeled and unlabeled data jointly. The training of an SSL model consist of using a small sample of labeled data D_L and extend its knowledge with a large portion of unlabeled data D_U. In this matter, we propose a semi-supervised segmentation model based on Teacher and Student training by following an iterative training process [14]. Our method first generates pseudo labels for unlabeled images using the teacher model and then feed it to the student model. After that, we update the student model each time. This process is repeated many times until the stopping criteria is met. The workflow of our model is presented in Fig. 1.

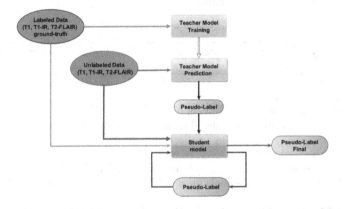

Fig. 1. The flow chart of the Teacher-Student approach.

2.1 Problem Formulation

Suppose we have a training dataset D. We divide it into two sub-dataset: One dataset is the labeled one $D_L = (X_i, Y_i)_{i=1}^{N}$, where X_i denotes the i-th image and Y_i denotes its ground truth(segmentation). The other dataset is the unlabeled one $D_U = (X_j)_{j=1}^{M}$, where X_j denotes the j-th image. here X and Y denote input images and a ground-truth segmentation label respectively. N and M are the number of sample in each dataset D_L and D_U respectively. $\hat{Y}$ denotes the predicted segmentation.

2.2   Model Training

Our teacher-student approach follows an iterative training architecture [14] presented in Algorithm 1.

Initially, we built our teacher model according to MMAN [13] a deep convolutional network. For the inference process, we start by training a teacher model in a supervised mode. After that, it will be used as pseudo-labels generator for the unlabeled dataset. Then we alternate the training of the student model and the pseudo-label generation procedures in an iterative manner to optimize the student model T times. The teacher model is trained in a supervised approach on a dataset D_L with its ground truth from the standard database (MRBrains Challenge [12]), which contains 3D images of brain tissue from multi-modal MRI. After training our Teacher model, we use it to predict pseudo-labels for the unlabeled dataset D_U. After generating the pseudo-labels $\hat{Y}_j$ for the unlabeled dataset D_U, the training set can be then enlarged by taking the union of both the labeled and the unlabeled dataset $D = \{D_L(X_i, Y_i) + D_U(X_j, \hat{Y}_j)\}$.

The student model is trained in supervised manner on this augmented dataset D the same way we train the teacher model. The overall training procedure is summarized in Fig. 1 and Algorithm 1.

Algorithm 1: The general algorithm for the training process of the Teacher-Student model

---

Inputs: Labeled dataset $D_L(X_i, Y_i)$, Unlabeled dataset $D_U(X_j)$
Outputs: SSL model M_S
Train the teacher model $M_T \leftarrow \text{Train}(D_L)$
Predict the pseudo-labels for unlabeled data $\hat{Y}_j \leftarrow \text{Predict } M_T(D_U)$
while *not stopping criterion* **do**
  $M_S \leftarrow$ a new model
  Train the student model $M_S \leftarrow \text{Train}(D_L \cup D_U(X_j, \hat{Y}_j))$
  Predict the new pseudo-labels for unlabeled data $\hat{Y}_j \leftarrow \text{Predict } M_S(D_U)$
end

While training the student model, we increased the attention and focus on the real manual labels in the labeled dataset, by placing higher weight values for the real labels in the training to increase confidence in the model training method.

2.3   Multi-Modality Aggregation Network (MMAN)

Multi-Modality Aggregation Network (MMAN) is a convolution neural network. It was proposed by Jingcong Li et al. [13]. MMAN is designed to be able for extracting multi-scale features from multi-modal MRI scan. It harnesses the complementary information extracted from multi-modality MRI of brain tissues for fast and accurate segmentation. The MMAN can segment the MRI of brain tissues into three different matters namely CSF, WM and GM, which is faster than many existing methods [13]. The MMAN (Fig. 2) is based on the Dilated-Inception block(DIB). This block is formulated using three expanding convolutional layers on parallel, one concatenation layer and one standard 1×1 convolutional layer. Figure 3 describes the inner structure of the DIB block. The use of dilated convolution benefit us by having a larger receptive field with same computation and memory costs while also preserving resolution without loss of coverage [15]. The MMAN is the base component of our model, which will serve as the teacher and the student model.

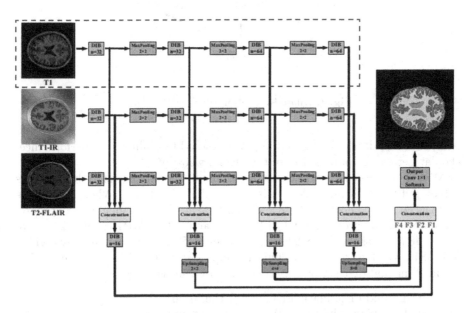

Fig. 2. The overall architecture of Multi-modality aggregation network (MMAN), that have many DIBs, Max-Pooling layers, Concatenate layers, Up-Sampling layers and an Output layer. The three images on the left are the multi-modality MRI(T1, T1-IR and T2-FLAIR) scan. The upper dashed box represent a sub-network for T1 modal. [13]

3   Experiment and Results

We devote this section to assess and discuss the achieved performance and efficiency of our proposed method.

3.1   Dataset

In this work, we used the **MRBrains Challenge** dataset which contains (20 subjects) in which 15 subjects are used as unlabeled data and (4) subjects are

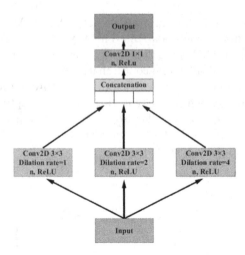

Fig. 3. Dilated-Inception block (DIB) [13].

used as labeled data. The last subject will be used to evaluate the performance. The dataset can be obtained from the official website[2].

MRI data were recorded by multiple (multi-modal) 3T MRI brain scans including T1, T1-IR and T2-FLAIR [12] as shown in Fig. 4. Where 4 subjects were placed as a labeled database and 15 other subjects as a unlabeled database. The data format in each modal (T1, T1-IR and T2-FLAIR) is $240 \times 240 \times 48$. That is 48 MRI (slices) 240×240 in size. These MRI have been manually divided into **CSF, WM and GM** by clinical neurologist and radiologist experts. The provided manual division by experts categorize the tissues into:

- 0: Background (The black area outside the brain)
- 1: Cerebro-Spinal Fluid CSF (including the ventricles)
- 2: Gray matter GM (gray matter cortical and basal ganglia)
- 3: White matter WM (including white matter lesions)

In our work, we consider 4 subjects with the special ground truth, which is the training group on which our proposed model(teacher) is trained in the first step. After which, 15 subjects are used as unlabeled data without the ground truth. Then, we will extract the pseudo-label from them using teacher model. Finally, our student model is trained on the sum of all 19 subjects with the sum of ground truth and pseudo-label as training data for our model. The last subject is left for evaluation purpose.

3.2 Evaluation Metrics

In medical image segmentation, several metrics are often used to evaluate the segmentation performance of different models. One of the most popular of these

---

[2] https://mrbrains13.isi.uu.nl.

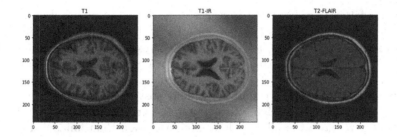

Fig. 4. The three MRI modals (T1, T1-IR and T2-FLAIR)

is the Dice coefficient (DC). It is a statistical tool that calculate the similarity between two sets of data. Dice coefficient is defined by the overlapping area between the predicted segmentation results and ground truth as follow [12]:

$$DC(G, S) = 2\frac{G \cap S}{G + S}.100\%$$ (1)

where S and G represent the predicted segmentation result and the ground truth respectively. DC is a metric that calculate overlapping area between the predicted segmentation result S and ground truth G. Expressed as percentages, the DC varies from 0% (total mismatch) to 100% (perfect match).

3.3   Experiment Setup

According to the construction of the MMAN model [13] as described, we rebuilt it and implement it from scratch. Initially, we trained it on a training dataset with the basic ground truth, which is (4 subjects) (192 slides of 240×240) out of (5 subjects) in the dataset. We left one subject to evaluate the model later. We used a teacher model(MMAN) to predict labels for a dataset of (15 subjects). In the next step, we combine the labeled dataset that the model was trained on and ground truth with the unlabeled dataset and the predicted pseudo-labels. Then we trained another model (based on the MMAN model) which is the student model on a whole dataset of (19 subjects). In our work, we trained (5 students) models in an iterative process that each time predicted pseudo-labels for the unlabeled dataset and combined it with the labeled dataset and trained the next model on it. This iterative method makes optimizing the results and reduces the loss function for the final model. The main objective function is to minimize the categorical cross-entropy loss function:

$$min_w L = -\sum_{i \in \Omega} \sum_c y_{ic} \log(p_{ic}) + (1 - y_{ic}) \log(1 - p_{ic})$$ (2)

where p_{ic} denote the predicted probability of cth class for pixel i while y_{ic} is the corresponding ground truth. Ω represent all of pixels in predicted segmentation result p and the corresponding ground truth y. If the pixel i belongs to the cth class, $y_{ic} = 1$ otherwise $y_{ic} = 0$.

Our model was implemented using the Python programming language. We used Tensor-Flow and Keras (Python libraries for building deep learning models) as tools to build our model and test it. The experiment was conducted in Google Colab, which is a cloud service that provides GPU-powered Notebooks. Based on empirical experiments, we found the optimal parameters for this model. In addition to parameters Optimizers (Adam), loss function (categorical cross-entropy) and provide the weight of each sample using (sample weights). During the training, we gave the labeled data more attention than unlabeled ones by giving them grater weight. According to the optimal results we have got in the empirical experiments, we set the weights for labeled data to 2 and unlabeled ones with 1. This will make the labeled data having more impact on the model during back-propagation process.

3.4   Result Discussion

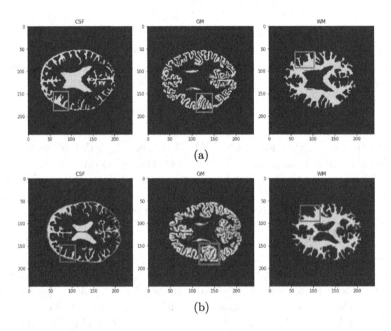

Fig. 5. (a) Ground Truth segmentation of MRI (WM, GM, CSF). (b) Predicted segmentation of MRI (WM, GM, CSF).

In evaluating the final model we used a new subject that had not been used previously in the training stage. The results are presented in Table 1 first row.

This method we followed in training gave better results, more accurate predictions 96.21% of mean of DC(including the background percentage). 82.63% and 86.25 for GM and WM respectively, where it achieved a reasonable result.

Table 1. Model evaluation results.

DC	Mean	GM	WM	CSF
TS MMAN [a]	96.21%	82.36%	86.25%	76.64%
Supervised MMAN [13][b]	–	86.15%	89.04%	84.52%

[a]Semi-supervised Teacher-Student MMAN (Ours)
[b]Fully supervised MMAN from the original paper

There are some minor decrease in performance that appear in pseudo-labels predictions of CSF (76.64%), which is the lowest of them all. We tried to improving this result by increasing the weights of the manual truth labels in student training.

The obtained performance based on the MRI scan data is presented for various modalities. Some segmentation examples are provided for ease of analysis Fig. 5a, 5b and 6, where most of noticeable errors occur in the frontier of the segmentation. As we can see from Figs. 5a and 5b, the red squares are some of the locations were our model could not be able to segment them accurately like green squares. These examples can help radiologist and neurologist to serve as Computer-Aided Segmentation System to gain accurate diagnosis and reduce consulting time.

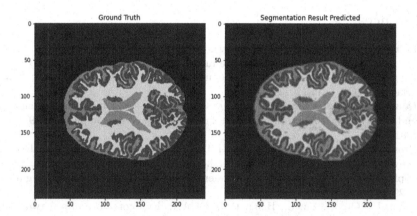

Fig. 6. Comparison of the correct and the predicted segmentation of MRI. (green, blue, yellow and Black) colors represent (CSF, GM, WM and background) respectively. (Color figure online)

The second row of Table 1 represent the results obtained from the fully supervised MMAN model presented in the work of author in [13]. As we can see from a comparison between the two model, the supervised MMAN has better performance than our TS MMAN with 3–4% difference regarding GM and WM. Whereas, there is a wide gap between the performances in CSF segmentation.

The difference in results is due to the learning strategies between the two models(supervised and semi-supervised approach), the size of labeled data in each model and the accuracy of pseudo-labeling of unlabeled data in our approach which will affect the training performance for the next student model; hence, the overall performance. Meanwhile, the CSF is located in the borderline and center of the brain, which overlaps with GM and WM and causes the low performance.

4   Conclusion

Throughout this paper, we designed a semi-supervised Teacher-Student models based on MMAN network. The process of training consist of an iterative adversarial cycle to improve pseudo-labeling accuracy. This models aims to segment multi-modal medical images of MRI brain tissue in an accurate manner without the need for huge amount of labeled data. The proposed method achieved a mean accuracy of DC of 96.21%. This model reached a comparable result to supervised models. but this result has some downside in frontier of the segmentation where there is some overlaps between classes. We aim to solve this problem in the next literature by using mini-patches of high quality scans to increase its segmentation capabilities and accuracy. Another approach is the employment of data augmentation and dynamic consistency/confidence losses which will give more attention to the labeled data.

References

1. Petrella, J.R., Coleman, R.E., Doraiswamy, P.M.: Neuroimaging and early diagnosis of Alzheimer disease: a look to the future. Radiology **226**(2), 315–336 (2003)
2. Giorgio, A., De Stefano, N.: Clinical use of brain volumetry. J. Magn. Reson. Imaging **37**(1), 1–14 (2013)
3. Dong, P., Wang, L., Lin, W., Shen, D., Guorong, W.: Scalable joint segmentation and registration framework for infant brain images. Neurocomputing **229**, 54–62 (2017)
4. Brosch, T., Tang, L.Y.W., Yoo, Y., Li, D.K.B., Traboulsee, A., Tam, R.: Deep 3d convolutional encoder networks with shortcuts for multiscale feature integration applied to multiple sclerosis lesion segmentation. IEEE Trans. Med. Imaging **35**(5), 1229–1239 (2016)
5. Chen, H., Qi, X., Yu, L., Heng, P.A.: DCAN: deep contour-aware networks for accurate gland segmentation. In: Proceedings of the IEEE Conference on Computer Vision and Pattern Recognition, pp. 2487–2496 (2016)
6. Ronneberger, O., Fischer, P., Brox, T.: U-net: convolutional networks for biomedical image segmentation. In: Navab, N., Hornegger, J., Wells, W.M., Frangi, A.F. (eds.) MICCAI 2015. LNCS, vol. 9351, pp. 234–241. Springer, Cham (2015). https://doi.org/10.1007/978-3-319-24574-4_28
7. Wright, R., et al.: Automatic quantification of normal cortical folding patterns from fetal brain MRI. NeuroImage **91**, 21–32 (2014)
8. Bai, W., et al.: Semi-supervised learning for network-based cardiac MR image segmentation. In: Descoteaux, M., Maier-Hein, L., Franz, A., Jannin, P., Collins, D.L., Duchesne, S. (eds.) MICCAI 2017. LNCS, vol. 10434, pp. 253–260. Springer, Cham (2017). https://doi.org/10.1007/978-3-319-66185-8_29

9. Chen, L.-C., et al.: Naive-student: leveraging semi-supervised learning in video sequences for urban scene segmentation. In: Vedaldi, A., Bischof, H., Brox, T., Frahm, J.-M. (eds.) ECCV 2020. LNCS, vol. 12354, pp. 695–714. Springer, Cham (2020). https://doi.org/10.1007/978-3-030-58545-7_40

10. Papandreou, G., Chen, L.C., Murphy, K.P., Yuille, A.L.: Weakly- and semi-supervised learning of a deep convolutional network for semantic image segmentation. In: Proceedings of the IEEE International Conference on Computer Vision (ICCV) (2015)

11. Wei, Y., Xiao, H., Shi, H., Jie, J., Feng, J., Huang, T.S.: Revisiting dilated convolution: a simple approach for weakly- and semi- supervised semantic segmentation (2018)

12. Mendrik, A.M., et al.: Mrbrains challenge: online evaluation framework for brain image segmentation in 3t MRI scans. Comput. Intell. Neurosci. **2015** (2015)

13. Li, J., Yu, Z.L., Gu, Z., Liu, H., Li, Y.: Mman: multi-modality aggregation network for brain segmentation from MR images. Neurocomputing **358**, 10–19 (2019)

14. Zhou, Y., Wang, Y., Tang, P., Shen, W., Fishman, E.K., Yuille, A.L.: Semi-supervised multi-organ segmentation via multi-planar co-training. arXiv preprint arXiv:1804.02586 (2018)

15. Yu, F., Koltun, V.: Multi-scale context aggregation by dilated convolutions. arXiv preprint arXiv:1511.07122 (2015)

Deep Layout Analysis of Multi-lingual and Composite Documents

Takwa Ben Aïcha Gader🆔 and Afef Kacem Echi[(✉)]🆔

National Superior School of Engineering, University of Tunis, LR: LATICE, Tunis, Tunisia
`takoua.benaicha@enis.tn, ff.kacem@gmail.com`

Abstract. It is crucial to accurately analyze the layout to convert document images to high-quality text. With the emergence of publicly available, large ground-truth datasets, deep-learning models have demonstrated their effectiveness in detecting and segmenting document layouts. This study presents a deep learning technique for document structure analysis, an important stage in the optical character recognition (OCR) system. Our method employs the YOLOv7 (Only Look Once version 7) model, a highly efficient and precise object detection model trained on the DocLayNet database. The trained YOLOv7 model quickly and efficiently identified and categorized different document components, such as caption, list item, text, table, section header, and picture. Regarding accuracy and efficiency, our evaluation demonstrates that the suggested method beats existing strategies, with strong generalization ability for diverse document layouts, text styles, and scripts.

Keywords: Document Layout Analysis · Object Detection · Deep Learning · Optical Character Recognition · Text Detection · YOLOV7

1 Introduction

The optical character recognition (OCR) method translates document images into machine-readable text. It has become increasingly important daily, allowing us to digitize and store documents quickly, search and index text, and extract information from scanned images or documents. OCR has many practical applications, such as document management, data entry, and text analysis. One of its primary benefits is its ability to save time and reduce the need for manual data entry. Thus it allows the preservation and easy access to essential documents. This can be useful in various domains, such as healthcare, finance, and legal industries, where a vast amount of data must be processed and stored. The quality of scanned documents and the presence of noise or distractions in images can impact the accuracy of OCR systems.

OCR technology faces a significant obstacle in layout analysis, which entails identifying and differentiating various sections of a document, such as text, images, tables, and graphics. This is important because it allows the OCR system

A. Bennour et al. (Eds.): ISPR 2023, CCIS 1940, pp. 76–89, 2024.
https://doi.org/10.1007/978-3-031-46335-8_7

to accurately extract the text from the document and ignore non-text elements. However, layout analysis can be challenging due to the variations in document layouts and image noise or distractions. For example, a document may contain multiple columns, different font sizes and styles, or overlapping text and graphics. These variations can make it difficult for an OCR system to identify and extract the text accurately. One solution for this issue is utilizing deep learning models to examine the document's arrangement and precisely recognize the written content. These models were used for image segmentation and other tasks like text detection and text recognition, giving challenging results. However, developing an effective layout analysis system remains an active area of research, as many challenges remain to be addressed.

This study proposes a method utilizing the powerful object detection model YOLOv7 for analyzing the structure of OCR documents and detecting text through deep learning techniques. The YOLOv7 is used to accurately and efficiently detect the scanned documents objects and to classify them into one of the six classes; picture, caption, list item, table, section header, and text. YOLOv7 is a powerful deep-learning model for object detection in images and videos proposed in 2022. A convolutional Neural Network (CNN) can be trained to locate and identify objects in an image or video by predicting bounding boxes around them and classifying them into given classes. This version of the YOLO model is one of the most recent updates, renowned for its speedy and precise object detection abilities. Real-time object detection and classification make it a valuable solution for a range of applications, including self-driving cars, security systems, and facial recognition. One of the critical features of YOLOv7 is its ability to make predictions on full images in a single forward pass rather than requiring a sliding window approach or region proposal method. This makes it faster and more efficient than some other object detection models.

The YOLOv7 model was trained using the DocLayNet database [6] in this work, and the results of our studies show that it beats existing techniques in terms of accuracy and efficiency. We tested it on Latin and Arabic documents of various layouts and text styles, and it shows good generalization ability. The proposed YOLOv7-based approach significantly improves OCR page layout analysis and text detection.

In the following sections, we will explore the proposed method in more detail. Firstly, we will examine related works in Sect. 2. Then, in Sect. 3, we will introduce the YOLOV7 model proposed for page layout analysis. Afterward, some experimental results are discussed. In Sect. 5, some conclusions and prospects are finally given.

2   State of the Art

Document layout analysis is an essential step in OCR systems, which aims to break down documents into different components such as text, headings, tables, figures, lists, and more. This process, known as document segmentation, helps structure the document's content and facilitates the transcription of the text by the OCR system. By separating the text from other document elements, the OCR system can focus on transcribing the text without being disturbed by formatting and layout features. Additionally, accurate and efficient segmentation of documents into their different components can improve the overall structure and readability of the transcribed text. Therefore, document segmentation is necessary to ensure the success of OCR systems.

One of the previous works aimed at solving the document analysis problem is the article proposed by [5] in 1993. The "Document Spectrum" (or "Docstrum"), a method of structural layout analysis based on unsupervised hierarchical clustering of page components, is described in this work. This method accurately measures the tilt, line spacing, and text block spacing, thus offering three significant advantages over many other methods: Independence from title angle, independence from varied text spacings, and the capacity to manage local regions with different text orientations in the same image are all desirable.

Several proposals have been made to solve the problem of layout analysis using deep learning methods. An example is the method proposed in 2012, a multiple-layer margin segmentation method for manuscripts with complex page layouts was presented in [1]. This technique employs straightforward and distinguishing characteristics obtained from the connected component level, resulting in the creation of resilient feature vectors. A multi-layer perceptron classifier is employed to classify connected components based on their pixel-level class. Following that, a voting process is used to refine the resulting segmentation and obtain the final categorization. The method was trained and tested on a dataset containing different complex margin layout formats, achieving a segmentation accuracy of 95%. In 2019, an innovative method was presented in [13] to identify logical structures in document images using visual and textual features. This technique uses two layers of LSTM recurrent neural networks to process the text in the identified zone, which is classified as a sequence of words, as well as the normalized position of each word concerning the page width and height. The labeled zones include abstract, title, author name, and affiliation. This approach achieved an overall accuracy of 96.21% on the publicly available MARG dataset. In 2022, a new deep learning approach [7] was presented to improve the understanding of document images. This system uses neural networks to convert physical documents into digital documents to extract the necessary information. The system relies on two autoencoder-decoder networks simultaneously segment text lines and non-textual components and identify non-textual elements. Tests conducted on RDCL2019 showed that this approach is more stable and adaptable to new formats than previous commercial and editorial systems. A recent study [11] proposed a new hybrid spatial attention network (HSCA-Net) to improve document layout analysis by enhancing feature extraction capability. The HSCA-Net comprises a spatial attention module, a channel attention module, and a lateral attention link. The performance of this network was evaluated on public datasets; PubLayNet, ICDAR-POD, and Article Regions.

In recent years, Object detection models like YOLO and SSD are useful in object recognition tasks; therefore, we decided to apply them to the problem of layout analysis in scanned texts. It should be noted that object detection involves determining the presence and location of objects in images or videos by answering two questions: where is the object located in the image, and what type of object is it? There are two primary categories in object detection: two-stage and one-stage approaches. In two-stage object detection, detection and recognition are separated, as with RCNN and Fast RCNN. In one-stage object detection, however, detection and recognition are merged into a single step, as with YOLO and Single-Shot-Detector (SSD). An example of a state-of-the-art method that uses object detection models for document analysis was presented in 2019 [2]. An innovative method based on YOLO was presented for table detection in scanned documents. This innovative approach includes adaptive adjustments, an anchor optimization strategy, and two post-processing methods, for example. The anchor optimization strategy uses k-means clustering to find more suitable table anchors. Additionally, post-processing techniques remove unnecessary white spaces and noisy page objects to improve the accuracy of table margins and UI scores. We also note the [4], which presented a deep neural network influenced by natural scene object detectors. The model was trained and evaluated using labeled samples from a large publicly available dataset, demonstrating the utility of object detectors in layout analysis. In 2022, a table detection model based on YOLO was proposed in [10]. The authors integrated an involution layer into the network backbone to improve its ability to learn the spatial layout features of tables. Additionally, they designed a simple pyramid network to enhance the model's efficiency, resulting in a 2.8% improvement in accuracy compared to YOLOv3 and a 1.25x increase in speed. Finally, we cite the work presented in [6], which introduced the DocLayNet database. This database contains 80863 pages that have been manually annotated from various sources, showcasing different layouts for the purpose of general document layout analysis. The authors trained multiple deep learning models, including YOLOV5, to highlight the diversity of the database.

We referenced the YOLOV7 as it is a one-stage object detector. It locates objects in an image by placing bounding boxes around them and identifies the objects' class or class probabilities, representing the object's name (as illustrated in Fig. 1). In this research, we use it to detect various document elements, such as text, titles, tables, figures, and lists. By training this model on the *DocLayNet* dataset of document images, we can automatically segment documents into their different components and extract meaningful text for further processing by the OCR system. This deep learning approach can significantly enhance the accuracy and efficiency of OCR systems.

3   Proposed Method

We present in this section the offered deep learning method for solving the issue of document layout analysis using the YOLOV7 model. This latter is a state-of-the-art object detection model well-suited for document layout analysis because it can accurately identify and classify different objects in images or videos using

Fig. 1. Example of image with detected and recognized components.

a single convolutional neural network (CNN). The YOLOv7 object detection model, proposed by Alexey Bochkovskiy et al. in [8], is a real-time object detection tool that uses single-stage detection. It was released in July 2022 and has been recognized as an exceptionally efficient and accurate object detector. It outperforms other deep learning models and YOLO versions in speed and accuracy, with performance ranging from 5 FPS to 160 FPS [8]. In [8], the authors discuss the improvements made to YOLOv7 and how it has set a new benchmark for object detection performance. They also describe the optimization techniques and strategies implemented to enhance the training and inference processes.

3.1   The YOLOv7 Architecture

In this study, we employed the original YOLOv7 architecture, in which image frames pass through the backbone network for feature extraction. These features are then combined and processed further in the neck of the network and passed on to the head, where the model makes predictions about the presence, location, and class of different objects contained in the image and generates bounding boxes around them (see Fig. 3 for an illustration of the used architecture). In the following, we detail the different key components of the YOLOV7 architecture:

- **Backbone Network**: the backbone network of the YOLOv7 model is a get the training lead head and auxiliary head labels simultaneously feature from the input image. The backbone selection is a key step, as it will improve object detection performance. In the original YOLOv7 paper, the authors used the E-ELAN deep CNN architecture (see Fig. 3) as the backbone network. It employs group convolution to enhance the cardinality of the incorporated features and then combines the features of different groups through a shuffle-and-merge method. This approach improves the feature representation and increases the efficiency of computation and usage of parameters (Fig. 2).

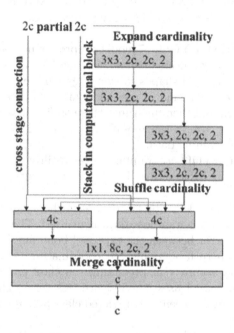

Fig. 2. The used extended ELAN (E-ELAN) [8].

- **Neck**: the neck of the YOLOV7 refers to the layers between the backbone network and the head layers (output layers). Its architecture is responsible for creating a feature pyramid, a set of feature maps at different scales that the head layers use to make predictions about the presence and location of objects in an image. The neck layers generally consist of convolutional, pooling, and up-sampling layers. For enhancement, authors of [8] used Attention Guided Feature Pyramid Network (FPN), feature refinement block (FRB), or Pyramid Attention Network (PAN).
- **Head**: In the context of the YOLOv7 architecture, the head refers to the output layers of the network that are responsible for making predictions about the presence, location, and type of objects within an image. The head layers take the features generated by the backbone and neck layers as input and use them to detect objects and produce the final bounding box coordinates

and class probabilities for each object detected in the image. The head layers are generally composed of a combination of convolutional layers and fully connected layers. We used the lead head guided label assigner as a head architecture. The training lead head and auxiliary head labels were obtained simultaneously through optimization using lead head prediction and ground truth. Otherwise, it generates the final predictions, and soft labels are derived from them. It's important to note that YOLOv7 has a Lead Head and an Auxiliary Head, which share the same loss function (see Fig. 3. Both heads are trained using the same soft labels. Still, the Lead Head is used to generate the final predictions, and the Auxiliary Head is used to refine the predictions of the Lead Head. This way, YOLOv7 can achieve high accuracy and real-time performance.

- **Anchors**: anchors in the YOLOv7 model are pre-defined bounding boxes that help the model identify objects in an image. These bounding boxes are chosen to have different sizes and shapes so that they can detect objects of various sizes and proportions. The model uses these anchor boxes as a starting point for object detection and then fine-tunes them to fit the objects in the image. These anchor boxes cover a range of object sizes and aspect ratios expected to be found in the training data.

- **Loss Function**: the YOLOv7 training process utilizes a composite loss function that includes multiple terms. The key components of this loss function include:
  - To enhance the model's ability to predict the exact location of objects in an image, a localization loss term is utilized. This term computes the mean squared error between the expected bounding box coordinates and the real ground-truth bounding box coordinates.
  - To improve its accuracy in predicting the class probabilities of objects in images, the model uses a confidence loss term. This term calculates the cross-entropy between the predicted class probabilities and the actual class labels.
  - To enhance the model's ability to anticipate the class labels of objects in an image, a classification loss term is computed by measuring the cross-entropy between the predicted and actual class labels.

The YOLOv7 loss function is a weighted sum of the localization, confidence, and classification Losses calculated for all the anchor boxes with an object present in the ground truth. When training, the YOLOV7 model aims to minimize this loss function by adapting the model's parameters to decrease the differences between the predicted bounding box coordinates, class probabilities, and class labels with ground-truth values. So, the model learns to recognize objects, their location, and their classification in the image.

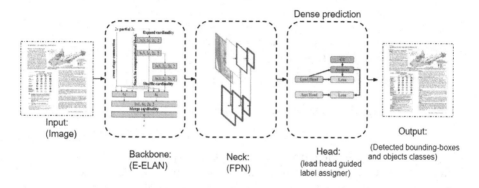

Fig. 3. The used YOLOV7 architecture.

4   Experimental Results

4.1   Used Database

We used the PubLaynet [12] database to test the YOLOV7 model for Document structure analysis. This dataset is utilized for analyzing document layout. It comprises of images of articles and research papers, along with annotations for diverse elements present on these pages, including text, lists, and figures. Over a million publicly available PDF articles from PubMed Central were mechanically matched with their XML representations to build the dataset [3]. In PubLayNet, there are several categories of document layout such as Caption, Page-header, Footnote, List-item, Page-footer, Formula, Picture, Table, Section-header, Text, and Title. However, we limit our training to six classes - caption, list item, picture, section header, table, and text - to enhance accuracy and precision.

Database Pre-processing. The database annotation is available in the COCO (JSON) format. The COCO format is a specific way of organizing labels and metadata for an image dataset using the JSON structure. However, to train our YOLOV7 model, the dataset must be in the YOLO format, so we reformated it in the acceptable format. In the YOLO format, each image in the dataset is associated with a single text file. If an image does not contain objects, it will not have a corresponding text file. The text file for each image contains rows of information, one row for each object presented in the image. For each row, we have five columns which indicate: Class-id denotes the class of the current object, x-center, and y-center represent the coordinate of the bounding box center, and a width and a height indicate the width and the height of the current bounding box (see Fig. 4 for an example).

To transform the database annotations from the COCO format to the YOLO format, we followed the following steps:

- Obtaining image-related information such as image_id, image_width, image_height, etc.
- Retrieve the annotations for a particular image using image_id.

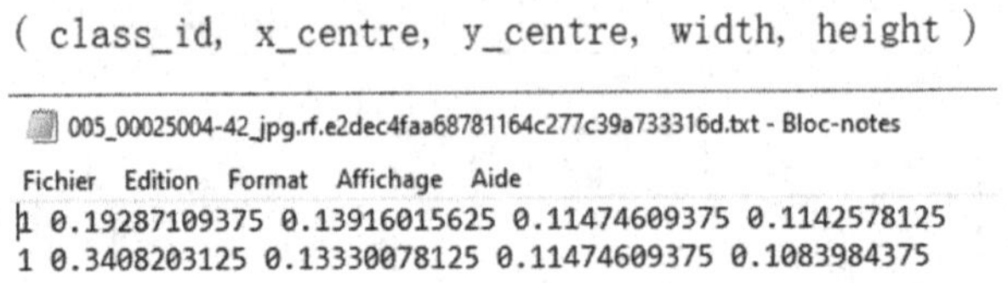

(class_id, x_centre, y_centre, width, height)

005_00025004-42_jpg.rf.e2dec4faa68781164c277c39a733316d.txt - Bloc-notes

Fichier   Edition   Format   Affichage   Aide

```
1 0.19287109375 0.13916015625 0.11474609375 0.1142578125
1 0.3408203125 0.13330078125 0.11474609375 0.1083984375
```

Fig. 4. Example of annotations in the YOLO format.

- Open a text file for the current image at the output path.
- Retrieve the bounding box properties for each object in the image.
- Save the retrieved annotations for the current image in a text file.
- Upon processing all the annotations for the current image, close the associated text file.
- Repeat the above steps for all images in the dataset.

Before using the converted dataset to train the YOLOV7 model, it was essential to ensure that it met the requirements and was converted correctly. To do this, we used the Roboflow online tool and checked sample images from the dataset (an example is shown in the Fig. 5).

4.2   Training

Instead of training our model from scratch, we opted to utilize transfer learning by starting with a generic COCO pre-trained checkpoint downloaded from [9]. Transfer learning is a methodology where a model previously trained for one task is adapted and refined to suit a different yet related task. This approach allowed us to take advantage of the knowledge and features learned by the pre-trained model and apply them to our document layout analysis task, saving us both labeled data and computational resources. As a result, we achieved better performance on the task with less cost.

For training the YOLOV model, we used multi-scale training, which involves training the model using images of different sizes. This helps the model to learn to detect objects of various sizes and in different contexts, which improves its ability to generalize and make more accurate predictions. Multi-scale training aims to prepare the model to detect objects in real-world images, regardless of their size. This can be done by randomly resizing input images during training or with different image scales like 32×32, 416×416, etc. We split the database into 70% for the training set, 20% for the validation, and 10% for the test and trained the model on the training dataset for 500 epochs using the training settings listed in Table 1.

The performance of the Yolov7 model on the training and validation sets is presented in Fig. 6, which displays three different types of losses: box loss, target loss, and classification loss. Box loss measures the algorithm's ability to identify the object's center and accurately predict its bounding box. Target loss evaluates the probability that an object will be found in a suggested zone of

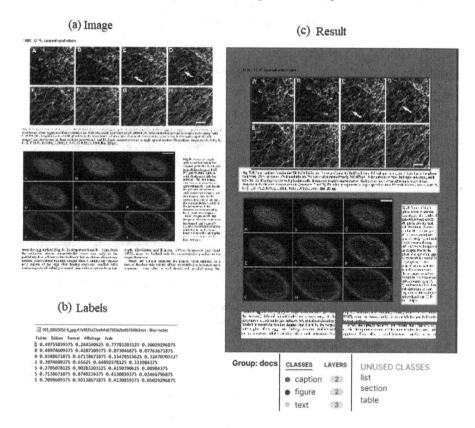

Fig. 5. Example of annotations in the YOLO format.

interest. Classification loss indicates the algorithm's capacity to classify an object correctly. The model's precision, recall, and average accuracy improved rapidly during training, as evidenced by the decreasing box, target, and classification loss values. The model improved precision, recall, and mAP after 200 epochs and achieved stability after 400.

4.3   Testing

Once the model was trained, it was utilized to analyze unseen images and video. Throughout the inference process, a confidence threshold of 0.1 was maintained. Figure 8 illustrates that the trained model could detect different objects presented on a document image and classify them to their accurate classes with prediction probabilities greater than 90% in most instances. We used the mAP (mean average precision) metric to evaluate the model's performance. An evaluation metric frequently utilized in object detection tasks. The mAP is calculated by calculating the Average Precision (AP) for each class and then averaging it over many classes (see Eq. 1). Precision (AP) is the proportion of true positive

Table 1. Training Parameters.

YOLOV7 model	hyperparameters
415 layers	$lr0 = 0.01$, $lrf = 0.1$, $momentum = 0.937$, $weight_{decay} = 0.0005$, $warmup_{epochs} = 3.0$, $warmup_{momentum} = 0.utilizeds_{lr} = 0.1$, $box = 0.05$, $cls = 0.3$, $cls_{pw} = 1.0$, $obj = 0.7$, $obj_{pw} = 1.0$, $iou_t = 0.2$, $anchor_t = 4.0$, $fl_{gamma} = 0.0$, $hsv_h = 0.015$, $hsv_s = 0.7$, $hsv_v = 0.4$, $degrees = 0.0$, $translate = 0.2$, $scale = 0.9$, $shear = 0.0$, $perspective = 0.0$, $flipud = 0.0$, $fliplr = 0.5$, $mosaic = 1.0$, $mixup = 0.15$, $copy_{paste} = 0.0$, $paste_{in} = 0.15$, $loss_{ota} = 1$

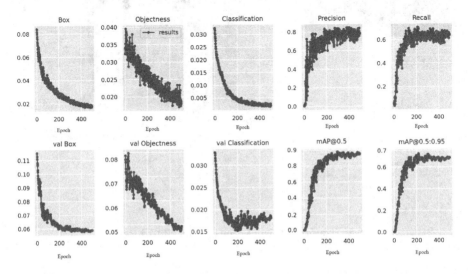

Fig. 6. The training and validation curves of box loss, objectness loss, classification loss, precision, recall, and mean average precision (mAP).

detections (correctly identified objects) among all positive detections (correct and incorrect).

$$mAP = \frac{1}{N} \sum_{i=1}^{N} AP_i \qquad (1)$$

The mean average precision (mAP) offers a consolidated value that accurately reflects the overall performance of the model. It is widely used in computer vision competitions and research to compare the performance of different object detection models. Table 2 shows our results compared to Mask R-CNN, Faster R-CNN, and YOLOv5 models on the test database based on the mAP@0.5–0.95(and the Fig. 7 represents the precision curves of the six classes on the test dataset. The following formula calculates the precision:

$$mAP = \frac{TP}{TP + FP};$$

(2)

where TP is the True positive and the FP is the False Positive.

The results show that the YOLOV7 model achieved the highest mAP compared to other models with respected gaps.

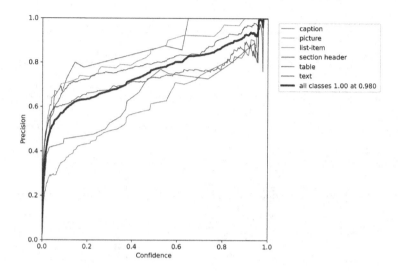

Fig. 7. Precision curves for the six classes for the test dataset.

Table 2. Prediction performance (mAP@0.5-0.95) of object detection models on DocLayNet test dataset.

Class	MRCNN [6]		FRCNN [6]	YOLO	
	R50	R101	R101	V5x6 [6]	**V7**
Caption	68.4	71.5	70.1	77.7	**80.9**
List item	81.2	80.8	81.0	86.2	**90.1**
Picture	71.7	72.7	72.0	77.1	**83.6**
Section-header	67.6	69.3	68.4	74.6	**84.0**
Table	82.2	82.9	82.2	86.3	**91.3**
Text	84.6	85.8	85.4	88.1	**92.7**
All	75,95	51,78	76,51	81,66	**87.1**

Fig. 8. Examples of the trained model's inferences on Arabic, Latin, and Chinese documents.

5   Conclusion and Prospects

We have introduced a new technique for analyzing document layouts, which involves utilizing the YOLOV7 model object detector. By training the model on the DocLayNet dataset, we could accurately detect and classify the various components of documents into one of eleven classes: caption, footnote, formula, list-item, page footer, page header, picture, section header, table, and title. This deep learning approach has proven to be efficient and effective, significantly improving the accuracy and efficiency of the base paper. The findings from this work indicate that object detection models can be utilized for document layout analysis, which presents a range of opportunities for automating document processing and analysis tasks. Future work could include further fine-tuning the model and incorporating additional image pre-processing techniques.

References

1. Bukhari, S.S., Breuel, T.M., Asi, A., El-Sana, J.: Layout analysis for Arabic historical document images using machine learning. In: 2012 International Conference on Frontiers in Handwriting Recognition, pp. 639–644. IEEE (2012)
2. Huang, Y., et al.: A yolo-based table detection method. In: 2019 International Conference on Document Analysis and Recognition (ICDAR), pp. 813–818. IEEE (2019)
3. Bethesda (MD): National Library of Medicine: PMC open access subset (2003). https://www.ncbi.nlm.nih.gov/pmc/tools/openftlist/. Accessed 15 Jan 2023
4. Minouei, M., Soheili, M.R., Stricker, D.: Document layout analysis with an enhanced object detector. In: 2021 5th International Conference on Pattern Recognition and Image Analysis (IPRIA), pp. 1–5. IEEE (2021)
5. O'Gorman, L.: The document spectrum for page layout analysis. IEEE Trans. Pattern Anal. Mach. Intell. **15**(11), 1162–1173 (1993)
6. Pfitzmann, B., Auer, C., Dolfi, M., Nassar, A.S., Staar, P.: Doclaynet: a large human-annotated dataset for document-layout segmentation. In: Proceedings of the 28th ACM SIGKDD Conference on Knowledge Discovery and Data Mining, pp. 3743–3751 (2022)
7. Tran, H.T., Nguyen, N.Q., Tran, T.A., Mai, X.T., Nguyen, Q.T.: A deep learning-based system for document layout analysis. In: 2022 The 6th International Conference on Machine Learning and Soft Computing, pp. 20–25 (2022)
8. Wang, C.Y., Bochkovskiy, A., Liao, H.Y.M.: Yolov7: trainable bag-of-freebies sets new state-of-the-art for real-time object detectors. arXiv preprint arXiv:2207.02696 (2022)
9. WongKinYiu: yolov7 (2022). https://github.com/WongKinYiu/yolov7/releases/download/v0.1/yolov7_training.pt. Accessed 18 Jan 2023
10. Zhang, D., Mao, R., Guo, R., Jiang, Y., Zhu, J.: Yolo-table: disclosure document table detection with involution. Int. J. Doc. Anal. Recognit. (IJDAR) **26**, 1–14 (2022)
11. Zhang, H., Xu, C., Shi, C., Bi, H., Li, Y., Mian, S.: HSCA-Net: a hybrid spatial-channel attention network in multiscale feature pyramid for document layout analysis. J. Artif. Intell. Technol. **3**(1), 10–17 (2023)
12. Zhong, X., Tang, J., Yepes, A.J.: Publaynet: largest dataset ever for document layout analysis. In: 2019 International Conference on Document Analysis and Recognition (ICDAR), pp. 1015–1022. IEEE (2019)
13. Zulfiqar, A., Ul-Hasan, A., Shafait, F.: Logical layout analysis using deep learning. In: 2019 Digital Image Computing: Techniques and Applications (DICTA), pp. 1–5. IEEE (2019)

A Novel Approach for Recognition and Classification of Hand Gesture Using Deep Convolution Neural Networks

Nourdine Herbaz[✉][iD], Hassan El Idrissi[iD], and Abdelmajid Badri[iD]

Laboratory of Electronics, Energy, Automation and Information Processing (LEEA&TI), Faculty of Sciences and Techniques Mohammedia, Hassan II University of Casablanca, Mohammedia, Morocco
herbaznourdine@gmail.com

Abstract. Sign language is a means of communication between individuals, whether they are hard of hearing or other people who do not speak the language of the host country. While it is a means of communication, few people know it and it has no universal patterns. For example, there is a set of signs unique to each country, resulting from the customs and traditions of each country. Currently, much research is being done in this area to solve the problems of sign language translation using computer vision and artificial intelligence.

The importance of this topic is due to the possibility of using Deep Convolutional Neural Networks (CNNs), embedded in deep learning technology, to recognize hand gestures in real time and to translate sign language.

This article presents two new datasets containing 54,049 of images of the ArSL-2018 (Arabic Sign Language) alphabet taken by more than 40 people, as well as 15,200 images in our personal datasets, categorized into 32 classes of standard Arabic characters, and classified, normalized, and detected using a VGGNet model and ResNet50. In this study, the success of two different training and testing exercises performed without fine-tuning and with fine-tuning enhancers was compared.

The optimized weights of each VGGNet layer were achieved as the network was pre-trained on two large datasets of alphabet sign language. In addition, the ResNet50 classifier was trained using 40 epochs and fine-tuned with 40 plus epochs to ensure optimal classification performance. The high accuracy levels achieved in comparison with other studies support the effectiveness of this approach. Specifically, the ArSL alphabets dataset achieved accuracies of 99,05%, 99,99%, and 98,50%, using VGG16, VGG19, and ResNet50 Models respectively, demonstrating the effectiveness of the proposed method for hand gesture recognition tasks.

Keywords: Deep Convolutional Neural Networks · Image Processing · Arabic Alphabets · Sign Language

© The Author(s), under exclusive license to Springer Nature Switzerland AG 2024
A. Bennour et al. (Eds.): ISPR 2023, CCIS 1940, pp. 90–105, 2024.
https://doi.org/10.1007/978-3-031-46335-8_8

1   Introduction

Gesture recognition plays an important role in human-computer interaction and immersive game technology due to its natural and user-friendly semantic expression. To use this technology, machines must detect them quickly and accurately so that users feel comfortable and ready to interact with the machines. Sign language recognition remains a difficult task due to their diversity, similarity of shapes of the hand gestures, and complexity of application scenarios.

The problem of Gesture recognition has various solutions in scientific papers. Existing approaches can generally be divided into two categories as shown in "Fig. 1":

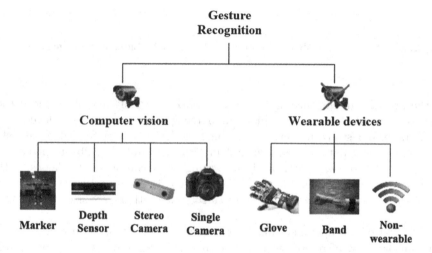

Fig. 1. Different types of gesture recognition [12].

- Recognition based on computer vision,
- Recognition based on wearable devices.

Among the techniques for solving the problem of gesture recognition, we find smart gloves [1,2]. This technology is used for the collection of data and their subsequent transmission. The data are hand movements such as rotational speed and angles, as well as finger angles transmitted by the flex sensors. The computer received data and performs recognition using a special algorithm, then translates it into speech or text. For example, Zhongda Sun et al. developed a haptic-feedback rings capable of augmented tactile-perception with multimodal sensing and feedback capabilities [3]. Shun et al. used a noncontact HMI system with high flexibility and robustness to control the robot based on hand-responsive IR structural colors [4].

Recently, to solve the problems of recognition of hand gestures, methods and solutions that do not require smart gloves appear more and more. Then, an analysis of existing gesture recognition methods using processing image based

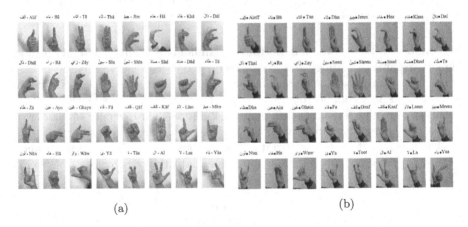

(a)   (b)

Fig. 2. a_ ArSL-2018 alphabets dataset [5], b_ Personal Arabic sign language alphabets.

in the artificial intelligence and neural networks. For this reason, we need the images of hand gestures to achieve high recognition rate, We employed two datasets for our study, namely the Arabic Sign Language (ArSL-2018) datasets [5] and personal datasets, as depicted in "Fig. 2a" and "Fig. 2b" respectively. The ArSL dataset was particularly intriguing, as it comprised 32 classes, with each class containing 54,049 images that were 64*64 pixels in size. On the other hand, the personal dataset consisted of 15,200 images that were sized to 224*224 pixels.

The rest of this document has been organized as follows: In Sect. 2, we start with an overview of related works; The methodology of our work is presented in Sect. 3; Sect. 4 presents the performance evaluation of our proposed model; Sect. 5 covers the experiments carried out, and the results recorded; and Sect. 6 concludes this work.

2   Related Work

In this section, we will detail literature review of various types of techniques implemented for classification and recognition of images of hand gesture using Machine/Deep learning. In recent years, a lot of researchers are working to collect and create a database of hand gesture for interpretation of sign language. Ghazanfar et al. [5] developed a full dataset of Arabic Sign Language (ArSL) to implement in the research related to hand gesture recognition, Barczak et al. [16] created a full dataset of standard American Sign Language (ASL) of hand gestures with 2425 images from 5 individuals etc.

Deep Convolutional Neural Networks (CNNs) are based on deep learning architecture and are the most widely used approach to classify and make a solution of different problems such as hand gesture recognition. In 2022, we [6] used the CNN method for training our personal dataset of hand gesture containing

1000 images to extract features map of hand gesture from images and translate them into text. In the same, Ghazanfar et al. [7] was designed a system to recognition Arabic Sign Language (ArSL) using Convolutional Neural networks (CNN). Alani and Cosma [9] established two models of CNN architecture, the first proposed ArSL-CNN model train and test accuracy increased from 98.80% to 96.59%, and the second proposed ArSL-CNN model with synthetic minority oversampling technique (SMOTE) test accuracy improved from 96.59% to 97.29%.

Most methods have been developed using histogram of oriented gradients (HOG) for feature extraction for hand gesture recognition. Adeyanju et al. [8] developed a System to recognize the American Sign Language using Canny Edge and Histogram of Oriented Gradient (HOG), and then adopted the K-Nearest Neighbour (K-NN) as a classifier to obtain hand contour edges.

By using the VGG16 feature extractor, Ewe et al. [10] used a deep CNN for hand gesture recognition. Authors trained the VGG16 model using very large dataset.

Ashish et al. [11] presented a Feature Extraction Techniques system to recognize of American Sign Language. In the preprocessing stage, the authors used many techniques as Support Vector Machines, Naïve Bayes, Logistic Regression, and K-Nearest Neighbours to find the information during feature extraction. Moreover, they compared it with the fateure extraction ORB.

Gnanapriya and Rahimunnisa [13] used modified UNET model to eliminates the performance of the background and segments in the images of hand gestures from the OUHANDS, HGR1 and NUS hand posture-II datasets. Their proposed method achieved an accuracy of 98.98%, 98.76% and 99.07% respectively.

3   Methodology

The proposed methodology consists of four major steps: Data classification, feature extraction, building Deep Convolution Neural Networks and Make a decision as shown in the diagram in Fig. 3.

In the first step, we used a big data of hand gesture to pre-processing, normalization and optimization for classifying our datasets.

The second step concerned to captured and take a new image for extracted the characteristics of hand gesture to recognize it.

In the third step, we used a Deep convolution Neural Network architecture to classify our approach.

Finally, in the fourth step, we trained and tested our system to make a decision and translate as a text.

3.1   Data Pre-processing

To evaluate our work efficiency, we used an Arabic Sign Language ArSL dataset, the dataset was collected from 40 participants of different ages [5] ..., and contains 54,049 images in 64 × 64 pixel size divided into 32 classes of Arabic hand

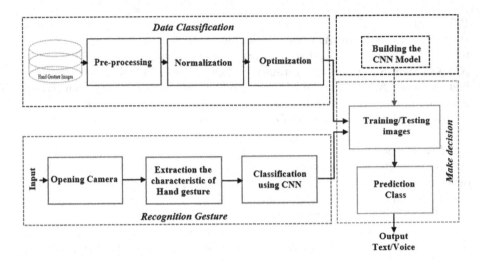

Fig. 3. Proposed diagrame for hand static gesture images.

Table 1. Settings for our studies.

Datasets	Name	Settings
ArSL-2018 [5]	Type of data	Images
	Image sizes	64*64
	Color space	Grayscale
	#Samples	54,049
	Nb. of classes	32
Personal datasets [19]	Type of data	Images
	Image sizes	224*224
	Color space	RGB
	#Samples	15,200
	Nb. of classes	32

gestures. A parameter settings used in our approach are shown in Table 1, and the full picture of Arabic alphabets is shown in Fig. 2.

3.2   Architecture System

In this proposed approach, a full VGGNet model has been implemented to efficiently extract information from an image that contains hand gestures, as shown in Fig. 4. Our VGG16 Model Fig. 4a, and VGG19 model Fig. 4b contained three main of layers: Convolution layers with additional activation layers named ReLU, Max-pooling layers, and Fully connected layers.

The VGG16 meaning visual Geometry Group with 13 layers of convolution and three Fully Connected layers were used in Fig. 4a, and 16 layers of convolu-

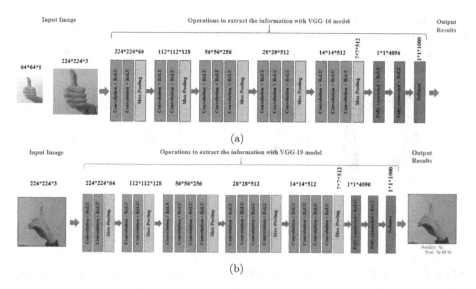

(a)

(b)

Fig. 4. a_ Architecture of VGG16 model for image classification, b_ Architecture of VGG19 model for image classification.

tion and three Fully Connected layers were used at VGG19 as shown in Fig. 4b. Before, we used a CNN architecture with five layers of convolution and three FC layers in our last work [6].

Deep Convolutional Neural Network (CNN) architecture is explained using simple diagrams in the following subsections:

Convolution Operation. The first operation is used to take the input images and process them. There are three elements in the 1st operation: Original image, Filters, and Output image (future map). In Fig. 5 we have an original image with a size of 5*5, each section we will convolued with a filter of size 3*3 for obtaining a new image with a reduced size named future map.

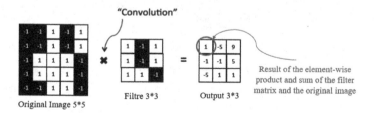

Fig. 5. Convolution operation.

Convolution+ReLU Operation. The second operation is used to transform the negative values to zero. In Fig. 6 we have a future map with negative values in red color, we will transform them into zero using the activation function ReLU.

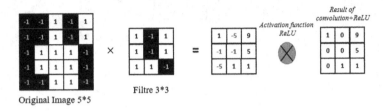

Fig. 6. Convolution + ReLU operation. (Color figure online)

Max-Pooling Operation. After the convolution layers, there is a Pooling layer. In our CNN model, we used Max-Pooling techniques, as shown in Fig. 7. The Max-Pooling is used to reduce the dimensionality and remove any noise existed in future map returned from previous convolutional layers. In our example shown in Fig. 7, we take the maximum value (Max-Pooling). In this sense, we eliminate a large number of unwanted pixels, and we keep a single value that contains the information. We reduced the size of input matrix 4*4 we move by a stride of two steps on all the pixels of the image with a filters of size 2*2. In this way, the image can be processed more efficiently.

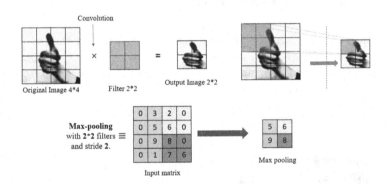

Fig. 7. Max-Pooling operation.

Fully Connected Layers Operation. Fully Connected Layer, these are the last layers of a CNN. There can be several fully connected layers. The last layer performs the classification task. As shown in this Fig. 8, after the first two layers the resulting image transform has a vector to access the neural network layers.

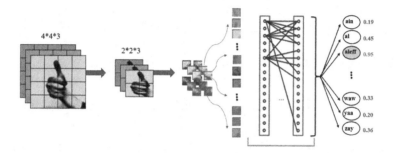

Fig. 8. Fully connected layers operation.

3.3 Flowchart

The proposed approach contains four phases - acquisition of data, pre-processing, deep learning, and decision. As shown in the flowchart in Fig. 9, a camera was used to capture the new images for pre-processing, classify and extract her characteristics to recognize the image's signature. The approach makes use the VGG16 model and CNN algorithm for feature detection techniques for all hand gestures.

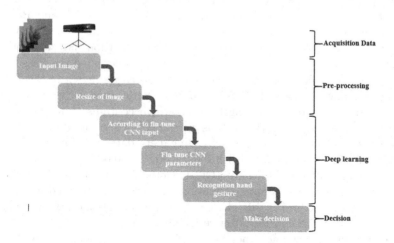

Fig. 9. Flowchart of Our Model Approach.

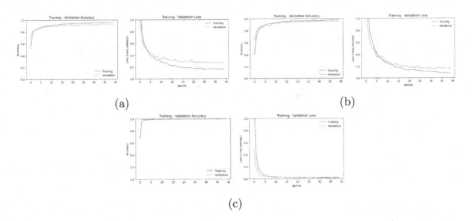

Fig. 10. (a) Training-Validation accuracy and loss curves for Resnet50 model without fin-tuned, (b) Training-Validation accuracy and loss curves for VGG16 model without fin-tuned, (c) Training-Validation accuracy and loss curves for VGG19 model without fin-tuned.

4   Performance Evaluation

In this approach, the performance of the Deep Convolution Neural Networks architectures created was evaluated using the VGGNet model, and ResNet50 Model with the dataset consisting of Arabic Sign Language (ArSL). We used 19200 images were divided into 9600 images for validation datasets and 9600 images for test dataset. To evaluate all these things, we created the classification report of the VGG16, VGG19, and ResNet models as we shown in the Table 2.

Once you have developed VGGNet and ResNet50 models, we can start to training and testing to visualizing a graph which contains training-validation accuracy and training-validation loss as we can see in the Fig. 10.

As a result of 40 epochs, the ArSL alphabets dataset achieved accuracies of 97.05%, 98.50%, and 99.50% with 0.02, 0.09, and 0.007 in the training loss using the ResNet50, VGG16, and VGG19 models, respectively. The training-validation accuracy and training-validation loss graphics obtained without the fine-tuning are shown in Fig. 10. These results are not satisfactory, so we used a fine-tuning of 40 more epochs. After re-training of our CNN model, as a result of 80 epochs we achieved: accuracies of 99.05%, 99.50%, and 99.99% with 0.006, 0.005, and 0.0007 in the training loss using the ResNet50, VGG16, and VGG19 models, respectively. The training-validation accuracy and training-validation loss graphics obtained with the fine-tuning are shown in Fig. 11.

5   Experiments and Results

In this study, our approach was codded using python 3 in Jupyter notebook platform. We used libraries of python which is an open source with Panda,

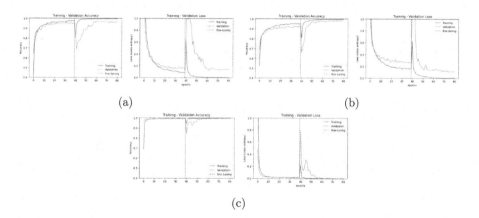

Fig. 11. (a) Training-Validation accuracy and loss curves for Resnet50 model with fin-tuned, (b) Training-Validation accuracy and loss curves for VGG16 model with fin-tuned, (c) Training-Validation accuracy and loss curves for VGG19 model with fin-tuned.

Seaborn, Scikit-learn, OpenCV, and TensorFlow. The model was carried out on hardware with Intel® Core i5-6300U CPU @ 2.40 GHz (4 CPUs), 2.5 GHz, and 8 GB of RAM.

At the end of all the training processes, finally our model is generated by 300 data by each class, the confusion matrix was created. In Fig. 12 the graph shows the confusion matrix.

The performance of our approach with VGG16 model is better than that of the CNN models we used in the last. The prediction accuracy of our approach on the ArSL alphabet dataset is shown in Fig. 13.

The performance of our methods is proving the simplicity and the high accuracy rate obtained, which means that all new images captured by camera are perfectly recognized With a very high learning rate. In Fig. 14, we show the Hand gestures with the very high accuracy rate the meaning of each sign.

To further demonstrate the performance of our approach model, we calculate each parameter's of classification (Precision, Recall, and F1- score), as shown in Table 2.

Table 3 presents a comparison of obtained results by our proposed approach with other previous techniques on the literature review of Hand Gesture Recognition.

We conducted extensive testing of our method on the two datasets mentioned previously and presented the results in Table 3. Other methods, such as those proposed by Alani et al. (2021) and Ghazanfar et al. (2020), utilized a CNN approach, where the recognition and detecting data was first encoded into an image before processing. Alani et al. method utilized a SMOTE oversampling technique to the class imbalance within the ArSL2018 dataset, while Ghazanfar et al. employed convolution neural network architecture to construct their network.

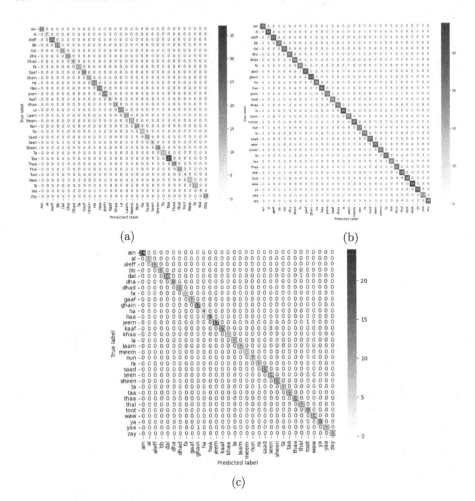

(a)

(b)

(c)

Fig. 12. (a) is representing Confusion Matrix VGG19, (b) is representing Confusion Matrix VGG16, (c) is representing Confusion Matrix Resnet50.

Barbhuiya et al. (2021) adopted an ensemble strategy based on modified AlexNet and modified VGG16 models for classification, using a VGG16 and AlexNet pre-trained, and SVM classifier for feature extraction. Ewe, E.L.R. et al. (2022) used a modified version of the VGG16 network that has been optimized for reduced computational complexity and memory usage, hence the name "Lightweight VGG16". After feature extraction, the classification is performed using a random forest classifier. The combination of these two techniques results in a more accurate and efficient hand gesture recognition system.

Khari el al. (2019) and Zihan et al. (2018) used a VGG19 deep network modules. Khari el al. (2019) used RGB and RGB-D as a feature concatenate

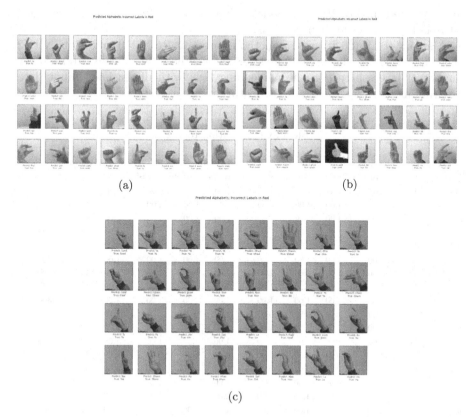

(a)   (b)

(c)

Fig. 13. (a) is representing Predicted Alphabets using Resnet50VGG19, (b) is representing Predicted Alphabets using VGG16, (c) is representing Predicted Alphabets using VGG19.

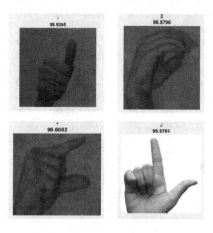

Fig. 14. Hand gestures with the very high accuracy rate.

layer for fine-tuned of VGG19 model. Zihan et al. (2018) used R-CNNs along with pooled features for convolutions.

Alnuaim et al. (2020) used a method based on ResNet50 and MobilNetV2 that relied solely on the embedding size. This method dynamically adjusted the number of layers based on the embedding size, with dimensional input data requiring more layers to extract the feature information.

Table 2. Classification report.

Alphabet	Classes	Resnet50 ArSL-2018			VGG16 ArSL-2018			VGG19 Personal Datasets			
		Precision	Recall	f1-score	Precision	Recall	f1-score	Precision	Recall	f1-score	Support
ain	0	1.00	1.00	1.00	1.00	1.00	1.00	1.00	1.00	1.00	300
al	1	1.00	1.00	1.00	1.00	1.00	1.00	1.00	1.00	1.00	300
aleff	2	1.00	1.00	1.00	1.00	1.00	1.00	1.00	1.00	1.00	300
bb	3	1.00	1.00	1.00	1.00	1.00	1.00	1.00	1.00	1.00	300
dal	4	1.00	0.94	0.97	1.00	0.97	0.99	1.00	1.00	1.00	300
dha	5	1.00	1.00	1.00	1.00	1.00	1.00	1.00	1.00	1.00	300
dhad	6	1.00	1.00	1.00	1.00	1.00	1.00	1.00	1.00	1.00	300
fa	7	1.00	1.00	1.00	1.00	1.00	1.00	1.00	1.00	1.00	300
gaaf	8	1.00	1.00	1.00	1.00	1.00	1.00	1.00	1.00	1.00	300
ghain	9	0.94	1.00	0.97	0.97	1.00	0.99	1.00	1.00	1.00	300
ha	10	1.00	1.00	1.00	1.00	1.00	1.00	1.00	1.00	1.00	300
haa	11	1.00	1.00	1.00	1.00	1.00	1.00	1.00	1.00	1.00	300
jeem	12	1.00	0.94	0.97	1.00	0.97	0.99	1.00	1.00	1.00	300
kaaf	13	1.00	0.83	0.91	1.00	0.88	0.95	1.00	1.00	1.00	300
khaa	14	1.00	1.00	1.00	1.00	1.00	1.00	1.00	1.00	1.00	300
la	15	1.00	1.00	1.00	1.00	1.00	1.00	1.00	1.00	1.00	300
laam	16	1.00	1.00	1.00	1.00	1.00	1.00	1.00	1.00	1.00	300
meem	17	1.00	1.00	1.00	1.00	1.00	1.00	1.00	1.00	1.00	300
nun	18	1.00	1.00	1.00	1.00	1.00	1.00	1.00	1.00	1.00	300
ra	19	0.92	1.00	0.96	0.98	1.00	0.99	1.00	1.00	1.00	300
saad	20	1.00	1.00	1.00	1.00	1.00	1.00	1.00	1.00	1.00	300
seen	21	0.82	1.00	0.90	1.00	1.00	1.00	1.00	1.00	1.00	300
sheen	22	1.00	1.00	1.00	1.00	1.00	1.00	1.00	1.00	1.00	300
ta	23	1.00	0.92	0.96	1.00	1.00	1.00	1.00	1.00	1.00	300
taa	24	1.00	1.00	1.00	1.00	1.00	1.00	1.00	1.00	1.00	300
thaa	25	1.00	1.00	1.00	1.00	1.00	1.00	1.00	1.00	1.00	300
thal	26	0.92	1.00	0.96	0.94	1.00	0.98	1.00	1.00	1.00	300
toot	27	0.94	1.00	0.97	1.00	1.00	1.00	1.00	1.00	1.00	300
waw	28	1.00	1.00	1.00	1.00	1.00	1.00	1.00	1.00	1.00	300
ya	29	1.00	1.00	1.00	1.00	1.00	1.00	1.00	1.00	1.00	300
yaa	30	1.00	0.92	0.96	1.00	0.94	0.99	1.00	1.00	1.00	300
zay	31	1.00	1.00	1.00	1.00	1.00	1.00	1.00	1.00	1.00	300
accuracy				0.98			0.99			1.00	9600
macro avg		0.98	0.99	0.98	0.99	0.99	0.99	1.00	1.00	1.00	9600
weighted avg		0.98	0.98	0.98	0.99	1.00	0.99	1.00	1.00	1.00	9600

Table 3. Comparison of obtained results by the proposed approach and other previous techniques.

Techniques	Authors	Datasets	#Samples	Accuracy(%)
CNN	Alani et al. [9]	ArSL-2018	54049	97.29
	Ghazanfar et al. [7]	ArSL-2018	54049	95.90
	Our approach [6]	Personal Data	1000	**98.70**
VGG16 model	Barbhuiya et al. [17]	ASL dataset	2520	99.82
	Ewe, E.L.R. et al. [10]	ASL dataset	87000	99.98
	Proposed approach	ArSL2018	54049	**99.05**
VGG19 model	Khari et al. [18]	ASL datasets	60000	94.80
	Zihan et al. [19]	ASL DAtaset	6000	98.70
	Proposed approach	Personal Data	15200	**99.30**
Resnet50	Alnuaim et al. [21]	ArSL2018	54049	97.50
	Proposed approach	ArSL2018	54049	**98.50**

6 Conclusion

Current techniques for solving gesture recognition have a number of drawbacks, such as low recognition speed, lag in response time, and poor performance in recognizing multiple targets or long-range targets in complex environments. In this study, we used the VGGNet, and ResNet model to train the ArSL dataset. We compared our results with existing systems that use the same datasets in the first step and with other use the different datasets in the second step. The proposed method achieved high levels of accuracy compared to other studies. Specifically, the ArSL alphabets dataset achieved accuracies of 99.05%, 99.99%, and 98.50% using the VGG16, VGG19, and ResNet50 models, respectively, indicating the effectiveness of the proposed method for hand gesture recognition. As expected, the model's performance outperformed other algorithms in existing systems with high accuracy. In future studies, We aspire to develop a system that can read sign language in continuous video, and furthermore, create an application mobile to recognize hand gestures to interpret sign language.

References

1. Wen, F., Zhang, Z., He, T., et al.: AI enabled sign language recognition and VR space bidirectional communication using triboelectric smart glove. Nat. Commun. **12**(5378) (2021). https://doi.org/10.1038/s41467-021-25637-w
2. Wang, Y., Tang, T., Xu, Y., et al.: All-weather, natural silent speech recognition via machine-learning-assisted tattoo-like electronics. NPJ Flex Electron. **5**(20) (2021). https://doi.org/10.1038/s41528-021-00119-7
3. Sun, Z., Zhu, M., Shan, X., et al.: Augmented tactile-perception and haptic-feedback rings as human-machine interfaces aiming for immersive interactions. Nat. Commun. **13**(5224) (2022). https://doi.org/10.1038/s41467-022-32745-8

4. An, S., Zhu, H., Guo, C., et al.: Noncontact human-machine interaction based on hand-responsive infrared structural color. Nat. Commun. **13**(1446) (2022). https://doi.org/10.1038/s41467-022-29197-5

5. Ghazanfar, L., Nazeeruddin, M., Jaafar, A., AlKhalaf, R., AlKhalaf, R.: ArASL: Arabic alphabets sign language dataset. Data Brief **23**(103777) (2019). https://doi.org/10.1016/j.dib.2019.103777

6. Herbaz, N., Elidrissi, H., Badri, A.: A Moroccan sign language recognition algorithm using a convolution neural network. J. ICT Stand. **3**(10) (2022). https://doi.org/10.13052/jicts2245-800X.1033

7. Ghazanfar, L., Nazeeruddin, M., Jaafar, A., AlKhalaf, R., AlKhalaf, R., Khan, M.A.: An automatic arabic sign language recognition system based on deep CNN: an assistive system for the deaf and hard of hearing. Int. J. Comput. Digit. Syst. **4**(90) (2020). https://doi.org/10.12785/ijcds/090418

8. Adeyanju, I.A., Bello, O.O., Azeez, M.A.: Development of an American sign language recognition system using canny edge and histogram of oriented gradient. Niger. J. Technol. Dev. **3**(19) (2022). https://doi.org/10.4314/njtd.v19i3.2

9. Alani, A.A., Cosma G.: ArSL-CNN a convolutional neural network for Arabic sign language gesture recognition. Indones. J. Electr. Eng. Comput. Sci. **2**(22) (2021). https://doi.org/10.11591/ijeecs.v22i2.pp1096-1107

10. Ewe, E.L.R., Lee, C.P., Kwek, L.C., Lim, K.M.: Hand gesture recognition via lightweight VGG16 and ensemble classifier. Appl. Sci. **12**(15) (2022). https://doi.org/10.3390/app12157643

11. Ashish, S., Anmol, M., Savitoj, S., Vasudev, A.: Hand gesture recognition using image processing and feature extraction techniques. Procedia Comput. Sci. **173**, 181–190 (2020). https://doi.org/10.1016/j.procs.2020.06.022

12. Gesture recognition in human-robot interaction: an overview. https://cnii-jest.ru/ru/science/publikatsii/53-raspoznavanie-zhestov-pri-vzaimodejstvii-cheloveka-i-robota-obzor. Accessed 25 Oct 2022

13. Gnanapriya, S., Rahimunnisa, K.: A hybrid deep learning model for real time hand gestures recognition. Intell. Autom. Soft Comput. **36**(1) (2023). https://doi.org/10.32604/iasc.2023.032832

14. Li, J., Li, C., Han, J., Shi, Y., Bian, G., Zhou, S.: Robust hand gesture recognition using HOG-9ULBP features and SVM model. Electronics **11**(988) (2022). https://doi.org/10.3390/electronics11070988

15. Wang, W., He, M., Wang, X., Ma, J., Song, H.: Medical gesture recognition method based on improved lightweight network. Appl. Sci. **12**(13) (2022). https://doi.org/10.3390/app12136414

16. Barczak, A.L.C., Reyes, N.H., Abastillas, M., Piccio, A., Susnjak, T.: A new 2D static hand gesture colour image dataset for ASL gestures. Res. Lett. Inf. Math. Sci. **15**, 12–20 (2011)

17. Barbhuiya, A.A., Karsh, R.K., Jain, R.: CNN based feature extraction and classification for sign language. Multimed. Tools Appl. **80**(2), 3051–3069 (2021). https://doi.org/10.1007/s11042-020-09829-y

18. Khari, M., Kumar Garg, A., González Crespo, R., Verdú, E.: Gesture recognition of RGB and RGB-D static images using convolutional neural networks. Int. J. Interact. Multimed. Artif. Intell. (2019). https://doi.org/10.9781/ijimai.2019.09.002

19. Zihan, N., Nong, S., Cheng, T.: Deep learning based hand gesture recognition in complex scenes. Pattern Recognit. Comput. Vis. (2018). https://doi.org/10.1117/12.2284977

20. Hand Gesture Image Dataset. https://universe.roboflow.com/hand-crpit/hand-gesture-hmohr/dataset/1. Accessed 08 Apr 2023
21. Alnuaim, A., Zakariah, M., Wesam, A.H., Tarazi, H., Tripathi, V., Enoch, T.A.: Human-computer interaction with hand gesture recognition using ResNet and MobileNet. Comput. Intell. Neurosci. (2022). https://doi.org/10.1155/2022/8777355

Breast Cancer Radiogenomics Data Generation Using Combined Generative Adversarial Networks GANs

Suzan Anwar[1]([✉]), Shereen Ali[2], Dalya Abdulla[3], Sam Davis Omekara[1], Salavador Mendiola[1], Kai Wright[1], and Saja Ataallah Muhammed[2,3]

[1] Philander Smith University, Little Rock, AR 72211, USA
sanwar@philander.edu
[2] Polytechnic University, Erbil, Iraq
[3] Salahaddin University, Erbil, Iraq

Abstract. Radiogenomics is a newly emerging field that integrates genomics with medical imaging data. The aim of this field is to elucidate the associations between gene expression data and imaging phenotypes, especially in cancer. However, radiogenomics is hindered by the expensive cost of genetic screening tests which leads to the unavailability of numerous big datasets for paired imaging and genetic data. Big data is crucial for training machine learning-based techniques for analysis relating to radiogenic studies. Currently, fake data is generated on only one of the two radiogenomic types; genomic or medical imaging data are generated separately. In this paper, we propose a deepfake approach implemented by combining two Generative Adversarial Networks (GANs) to create fake image data that are hardly differentiable from the original to improve Breast cancer diagnosis. To evaluate the model, a survey is developed and distributed among the participants to measure their ability to differentiate the original from the Deepfakes images. The results showed that the model-generated fake images cannot be distinguished from the authentic images and are relatively satisfying using the PyTorch framework.

Keywords: Radiogenomics · Generative Adversarial Networks · Deepfake · Breast Cancer

1 Introduction

Radiogenomics is a promising field with many valuable potentials, such as shedding light on underlying disease mechanisms, survival estimation, and treatment response prediction. Existing approaches for generating fake data can create either genomic data or medical imaging separately. These approaches ignore the associations between the disease's genetic mutations and their effect on the structure and function of human tissues and organs. Generating big fake genomic data with existing realistic imaging data will have a broad scientific impact as it will improve cancer diagnosis prediction access to life's domains.

A. Bennour et al. (Eds.): ISPR 2023, CCIS 1940, pp. 106–119, 2024.
https://doi.org/10.1007/978-3-031-46335-8_9

Radiomics and radiogenomics provide a comprehensive view of tumors through image data. In radiomics analysis, large amounts of quantitative data are extracted from medical images. The data obtained is then combined with clinical and patient data to create mineable shared databases. Radiogenomics expands upon radiomics as it further combines genetic and radiomic data [1–3]. Genetic testing is unavailable to all patients due to its cost, invasiveness, and time commitment. This emphasizes the need for radiogenomic, a discipline that offers a workable substitute since it might be crucial in delivering imaging surrogates that are related to genetic expression. [4–6]. Imaging surrogates can be utilized to tailor patients' treatment options, predict treatment response and the possibility of early metastases [7–9]. While a few studies highlighted the associations between gene expression data and imaging phenotypes, it is still being determined if this broadly applies to generate an extensive radiogenomics dataset, representing a critical knowledge gap that this work aims to close. We will use a deepfake-based approach that generates either or both types of radiogenomic data. The use of software applications to create high-quality altered videos has contributed to the widespread popularity of deepfakes. These applications are easy to use with users ranging in computer proficiency from experts to beginners. Deep learning techniques are utilized when developing these software. Complex and High-Dimensional data representation is a popular use case for deep learning. For example, deep auto-encoders, a variant of deep networks, have been widely employed for image compression and dimension reduction tasks [10–12]. A potent deep learning technique utilized for entire-face synthesis is StyleGAN [13], where the framework uses the mapping of points in latent space to an immediate latent space to regulate the style output at each stage of the creation process. StyleGAN can produce extremely photorealistic and high-quality images of faces after being trained on the FFHQ dataset 1, with only minor elements potentially pointing out the forgery. Karras et al. improved the generator (including redesigned normalization, multi-resolution, and regularization algorithms) and proposed StyleGAN2 to address these flaws [14] to obtain highly realistic faces. In order to develop new Deepfake detection techniques, different traces left by the generative architecture used in synthetic multimedia data can be retrieved and examined. Zhang et al. [15] studied the artifacts that the GAN pipelines' up-sampler introduced in the frequency domain. The spectrum-based classifier's results considerably improves the generalization ability and produce remarkable results in the binary classification test between real and altered images.

Guarnera et al. [16, 17] proposed a useful idea: each GAN architecture has a distinctive pattern that reflects its neural structure and how it generates images, like a fingerprint of the creation process. A Cycle-GAN network is proposed to translate unpaid images using two GAN networks while a cycle-consistency loss is included to map the inputs with the outputs images [18]. The temporal artifacts examined include detecting variations in behavior, measuring physiological indicators, ensuring coherence, and aligning video frames. Some data-driven approaches identify modifications through categorization or anomaly identification rather than focusing on a particular artifact [19]. The Deep Convolutional GAN (DCGAN) is a type of convolutional GAN that uses specific architectural constraints to improve the stability of the training process.. They also demonstrate significant progress in image-to-image translation using a generative adversarial network (GAN) framework [20]. Deep Neural Networks (DNNs) in partic-ular have become prevalent as new AI technologies, and this has dramatically increased the number of fraudulent or altered images, videos, and audio. Although the alteration of multimedia has been done in the past, cloning human voices and creating compelling fake human face images or videos has become much simpler and quicker today. The most popular deep learning-based generation techniques use Generative Adversarial Networks (GANs), Variational Autoencoders (VAEs), and Autoencoders (AEs). These techniques integrate or superimpose a human face image from a source onto an image from a tar-get [21]. It is demonstrated how a complex Long Short-Time Memory (LSTM) can be utilized to examine the authenticity of media files such as movies and photographs, as well as a temporal-aware system to automatically detect deepfake films [22]. Deepfake videos on the internet differ significantly from deepfake videos in research community datasets. The differences are attributed to content, and generation methods, raising new challenges for detecting deepfake videos. Detection methods were also considered poor performance, which can be attributed to racial biases [23]. The use of technological solutions such as automated systems for content validation and deepfake detection is a constantly evolving field of cybersecurity. Although deepfakes can have positive appli-cations, they also pose significant threats to society. To combat the negative impact of deepfakes, there are four main methods: legislation and policy regulation, corporate poli-cies and voluntary action, education and training, and anti-deepfake technology. Legal action should be taken against those who create harmful deepfakes, and it may be par-ticularly effective in cases involving foreign states [24]. Security practices implemented in healthcare systems are lacking, making them vulnerable to attacks compromising medical data availability and integrity. It is advised that more algorithms will be imple-mented to detect fraud better as advances in deep learning continue. Jekyll is designed to translate images from X to Y, injecting a new disease condition while preserving the patient's identity. Jekyll could also be used for disease removal [25].

Some synthetic ECGs were created using multiple publicly available accurate ECGs as data, not just a specific individual. These fake ECGs can then be shared with other researchers since they mimic accurate ECGs and do not include sensitive content that can be used to verify a patient's identity. Thus, concluding that no privacy issues were raised while still aiding in research. The synthetic ECGs were put through the GE MUSE system to ensure they were realistic using a deepfake approach. The results revealed that 81.3% of the 150,000 deepfake ECGs were deemed as "normal" while the rest were identified as "non-normal"; none of the fake ECGs were classified as invalid [26].

The role that deep learning has in Deepfake technology, the different ways of creating deepfakes, and the different strategies used to detect deepfakes are discussed in detail. Fake image detection is divided into deep learning methods (neural networks are used) and handcrafted techniques (neural networks are not used). Furthermore, fake video detection is similar in that it looks closely at the multiple frames of a video by making comparisons on a structure-by-frame basis. Fake video detection can be divided into temporal feature detection and visual artifact detection. The former looks at the fake video as a whole using recurrent network models; the latter focuses more on single frames in a fake video that may have suspicious tampering [27].

In this work, the deepfake approach depends mainly on combined GANs (Generative Adversarial Networks), which are trained using realistic genomic and imaging data. GANs architecture has two essential elements: the generator and discriminator networks. The generator network generates fake data, and the discriminator differentiates between the fake data and the realistic data and sends its feedback to the generator network, which uses this feedback to increase the outputs' quality and bring them closer to the actual data. The process continues until the discriminator cannot differentiate between authentic and original data. The contribution of this paper can be listed as follow:

1. This research will lay the groundwork for creating a target prediction database, empowering researchers to identify the underlying disease mechanism.
2. Generating genetic and imaging data for Breast cancer disease dataset for training using Machine Learning ML techniques.
3. Using a deepfake approach that leverages powerful deep learning tools to generate altered data that are hardly differentiable from original data to improve Breast cancer diagnosis.

2   Methodology

A major subfield of artificial intelligence is Machine learning. Machine learning studies computer algorithms that develop on their own through experiences. Machine learning enables systems to learn from data and make predictions or decisions without explicitly programmed. It can be applied to various tasks such as image recognition, speech transcription, search relevance, and personalization of content or services. The aforementioned tasks rely on deep learning techniques. Deep Learning is a type of AI that simulates the human brain's way of processing data. Some examples are voice recognition, object recognition and detection, language translation, and decision making. Deep learning is independent since it does not need human supervision and can learn from unstructured and unlabeled data.

Face detection was a research topic of interest from the late 90s to the early 2000s, due to its potential applications in security and the military. This problem has been solved in the past two decades, leading to face detection technology being easily accessible through open-source libraries in most high-level programming languages. Some of the most popular Python face-detection libraries are OpenCV and face recognition. These libraries enable various apps to advance to the level of being able to swap faces in images. Apps like FaceApp allow users to visualize themselves with another person's body or swap faces with famous individuals, such as celebrities or politicians. GANs learn differently from other kinds of neural networks. GANs use two neural networks called a Generator and a Discriminator, which compete with each other to produce the desired outcome. The Generator's job is to create realistic but fake images, while the Discriminator is trasked with distinguishing between real and the Generator's fake images. What is remarkable is that if both neural networks perform well enough, the results are images that are indistinguishable from real photos. Generative Adversarial Networks have been very successful since they were proposed in 2014 by Ian J. Goodfellow. Adversarial networks were developed after researchers observed that conventional neural networks could be manipulated into misclassifying data by adding even a small amount of noise.

Interestingly, a model often exhibits greater confidence in an incorrect prediction when noise is added, as opposed to when it makes a correct prediction. This behavior occurs because many machine learning models learn from a limited amount of data, which can lead to overfitting and other drawbacks. Moreover, the mapping between the input and the output is almost linear. Although it may appear that the boundaries of separation between the different classes are linear, in reality, they are made of linearities. Even a minor change in a point in the feature space might result in the misclassification of data.

This work uses an advanced deep learning algorithm to build an artificial intelligence model for learning from actual data to generate fake genetic and imaging data using GANs. We used the Python programming language, and its scientific packages, such as PyTorch, to build the GAN model and the convolutional network layers, implement the algorithm, and deploy the algorithm on the Amazon servers to guarantee better performance in terms of speed and accuracy. Following, we will explain the methodology for our work by listing all the used models, technologies, and approaches, along with the mathematical algorithm. Finally, the results and conclusion are explained with the necessary figures in the following sections.

2.1 Generative Advance Network GAN

GAN is a specific type of deep learning based on convolutional neural networks (CNN). Developed by Goodfellow [28], GANs work by generating new data based on a training set. For example, if a GAN is trained on a set of photographs, it can create new photos that resemble the ones in the training set. These newly created photos are often called Deepfakes. CNN assigns learnable weights and biases to each aspect of an object by taking an input image and distinguishing one from another. Likewise, the GAN network builds two neural networks: discriminator and generator. Together, these networks differentiate between the input samples and the created information. These Deepfakes are images created by deep learning to construct fabricated media in which a person's features in an existing photo or video are replaced with another person's features. The rise of Deepfakes is a growing societal concern as it becomes increasingly difficult to distinguish between real and fake media. These manipulated images and videos have been used to deceive the public and spread fake news. Because photos generated by the machine are flawless these days, it has become difficult to tell apart model-created images from the originals. Notable examples include the video of Barack Obama insulting Donald Trump, Mark Zuckerberg boasting about having access to billions of people's stolen data, and Jon Snow's emotional apology for the disappointing ending to Game of Thrones. Deepfake technology has also been used to create pornographic images of celebrities. In September 2019, AI company Deeptrace discovered 15,000 deepfake videos online, and this number continues to rise. The technology has become so accessible that even inexperienced individuals can create and share deepfakes. In addition to images and videos, audio can also be manipulated using the Deepfake method to create "voice skins" or "voice clones" of public figures. This poses a threat to the credibility of audio recordings and raises concerns about the spread of fake information. Recent frauds purportedly take advantage of speech recordings from Facebook, YouTube, and WhatsApp. In their work, scientists and researchers discussed the potential threat posed by special effects studios to video and image manipulation. To generate face-swap photos or films, a few simple procedures are required. First, thousands of people's faces will be used as input to the encoder, one of the AI algorithms.

Two faces' features will be learned by the encoder, who will then reduce them to their essential components and compress the photos. A decoder, the second AI algorithm, will be taught to extract the original from the created images, on the other hand. The faces of the first and second people will be recovered using two different decoders, respectively. The method feeds the encoded images into the "wrong" decoder to carry out the face swap. For instance, the decoder trained on the image of the second person is fed a compressed image of the first person's image. The second person's face is then developed by the decoder using the first person's facial orientation. This needs to be generated for deepfake videos on each frame. On a typical computer, it is challenging to create a convincing deepfake. High-end workstations with powerful graphics cards or, better yet, cloud computing are used to make the majority of fake videos. The processing time is reduced with this method from days to weeks to hours. But editing finished videos to remove visual flaws also requires skill. There are many deepfake programs available right now to create deepfakes of photos, videos, or data. Businesses create deep fakes for consumers and process them online. There are hundreds of smartphone apps that can be downloaded and used to add users' faces to TV lists for training. Deepfake picture and video producers and distributors may be in violation of the law. A deepfake may violate copyright, violate data protection laws, or be libelous if it subjected the victim to mockery, depending on the material. Without the owner's permission, sharing photographs is illegal and can result in up to a year in prison for the offender.

A common method for producing deepfakes from a dataset is GAN. Figure 1 displays a block diagram of a typical GAN network. The fundamental module of the GAN network consists of two networks: the generator $G(x)$ and the discriminator decoder $D(x)$ networks. The generator produces an imitation image that resembles those in the training set. This method is one of the machine learning approaches known as a deep fake, even though the discriminator distinguishes the real image from the deepfakes image produced by the generator. Using video, audio, or image files as training material, both the discriminator and the generator simultaneously learn new information. During the training phase, a data collection X with a large number of real photos is employed, distributed according to Pdata. The objective of the generator G is to produce images $G(z)$ that are similar to real images x, where z denotes noise signals under a distribution of Pz. Discriminator D attempts to separate real photos x from images created by G. D stands for the likelihood that an input image is recognized by D as being real rather than a fake image created by G.(input). D(input) obviously [1, 1]. While D has been trained to increase the likelihood that D correctly classifies both real photos and fake images created by G, G is being trained to reduce the possibility that D would identify its outputs as false images, or to minimize 1 D(G(z)). The following value function [28] can be used to define this two-player minimax game between D and G:

$$\text{Min}_G\text{max}_D V\ (D,\ G)\ = E_{X \sim P_{data}(X)}\left[loglogD(X)\right] + E_{Z \sim P_z(Z)}[D(G(Z)))] \quad (1)$$

After sufficient training, the generator ought to be able to produce incredibly life-like images from noise signals z. In the meanwhile, it will become simpler for the discriminator to distinguish between actual and fake photographs.

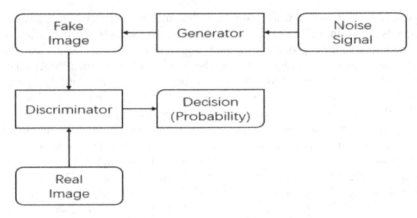

Fig. 1. A block diagram of GAN [28]

2.2 Deep Convolution GAN DCGAN

A GAN network that has architectural topological constraints applied to it to provide training stability is called Deep Convolution GAN DCGAN [29]. The DCGAN consists of CNNs with different layers for both the generator and the discriminator. DCGANs are used for style transfer; for instance, the network will generate a set of handbags that has the same style as an inputted set of shoes. In this work, we used two DCGAN generators and two discriminators.

2.3 Conditional GAN CGAN

The conditional GANs generate better-quality images by using extra label information. It has the ability to manage the look of the fake images and exploit the input information to the model and produce better images. This network allows the use of additional information, such as class labels, by setting some conditions features. The actual labels will be passed to the CGAN during the training phase to learn the differences between the labels.

2.4 Stacked GAN StackGAN

The main purpose of using the StackGAN is to generate images from the provided text description. Another usage is image-to-image translations by using sketches produced from objects' real images. For instance, a StackGAN can generate a fake image of a running boy from statements that describe the action and image. It consists of two GANs that are stacked together, producing a high-resolution image network.

2.5 Information Maximizing InfoGAN

InfoGAN is known as an information-theoretic extension of the GAN. It uses an unsupervised learning approach to learn disentangled representations. InfoGANs are utilized when the dataset is complex, not labeled to train a cGAN, and to see the most significant features of the inputted images. The InfoGAN increases the mutual features between a small subset of noise variables of the GAN and the observations.

2.6   Wasserstein WGAN

WGAN is an alternative to traditional GAN training. The discriminator in the WGAN provides a better learning signal to the generator, and more stable training is allowed when the generator learning distribution is in high dimensional spaces. The cost function is modified by the WGANs to incorporate a Wasserstein distance. They employ cost functions that are related to the caliber of the produced images.

2.7   Discover Cross-Domain Relations with Generative Adversarial Networks Disco GANs

DiscoGANs are basically used to generate images in a specific domain with a specific style and pattern by resembling the images in a different domain in both pattern and style. The GAN was learned without pairing images from different domains during the training phase, which reduced the time-consuming for image generation.

2.8   Our Proposed DCGANs

In this work, two different sets of images are used to feed each DCGAN. The generated fake images from noise signals for one DCGAN are difficult to distinguish from the other. Each discriminator aims to differentiate between the genuine image and the fake image produced by each generator inside its image domain, as shown in Fig. 2, which shows the pipeline of our work. The first stage of our proposed work structure is accepting two different inputs, images X and Y. In the processing stage, the region of interest in the images will be detected, cropped, and aligned in each inputted image. In the training stage, the fake images generated from noise signals from set X input images should be similar to domain Y images. At the same time, generated images from set Y input images should be similar to domain X images.

Our generators and dissertators networks consist of eight layers; in each layer, batch normalization and Leaky ReLu were done, and the tanh activation function was returned. Both networks are combined into one model, which is used to train the weight by the generators, then use the output to calculate the error by the discriminators.

We calculated the two losses to ensure the mapping of the generated images from each DCGAN and their input images. When the genuine images are utilized as inputs to the generators, this is intended to promote the loss to be close to an identity mapping. The two generator losses are combined to get a generator loss for the whole network, and the two discriminator losses are combined to get the disseminator loss for the whole network.

$$L_{GAN} = E_{x \sim P_{data}(x)}\Big[(1 - D_Y(G(x)))^2\Big] + E_{y \sim P_{data}(y)}\Big[(1 - D_X(G(y)))^2\Big] \quad (2)$$

$$L_D = E_{x \sim P_{data}(x)}\left[D_Y(G(x))^2\right]$$
$$+E_{y \sim P_{data}(y)}\left[D_X(F(y))^2\right]$$
$$+E_{x \sim P_{data}(x)}\left[1 - D_X(x))^2\right] \qquad (3)$$
$$+E_{y \sim P_{data}(y)}\left[1 - D_Y(y))^2\right]$$

where X and Y are two domains,
x, and y are different sets of input images,
G is the generator of images in domain X
F is the generator of images in domain Y
D is the discriminator of the GAN

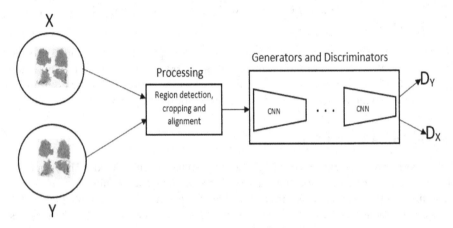

Fig. 2. The overall pipeline of our work

The developed images are preserved, and a fresh dataset demonstrates how our effort progresses from creating random noise to producing increasingly lifelike images.

3 Results

The deepfake module is trained using the selected TCGA radio genomics dataset. The collected data is labeled to know the true outcome from the prediction of our model and to compare the prediction from our models with the ground truth.

3.1 Cancer Genome Atlas Dataset

We gathered more than 2.5 petabytes of genomic, epigenomic, transcriptomic, and proteomic data using the Cancer Genome Atlas (TCGA), which has already improved our capacity to detect, treat, and prevent cancer. [30]. Figure 3 shows samples of the TCGA dataset. Figure 4 shows two different TCGA dataset breast tumor positions.

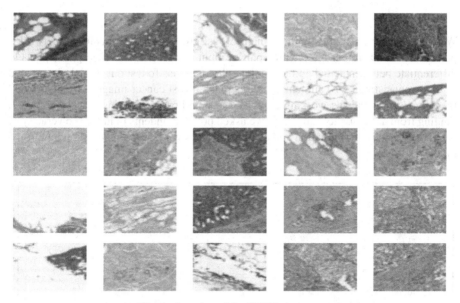

Fig. 3. Samples of the TCGA dataset.

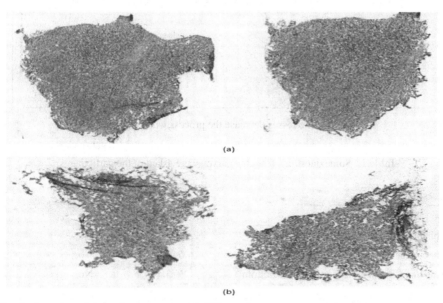

Fig. 4. Samples of the TCGA dataset with different breast tumor positions (a) The top section of the Breast with 92% of tumor nuclei. (b) The bottom section of the Breast with 80% of tumor nuclei.

3.2 Evaluation

We used perceptual studies with the help of test group data to evaluate the network's performance. When presented with a pair of input and output, we asked participants to differentiate between actual and fraudulent photographs. To test our model, we created a survey consisting of 10 authentic and deepfake breast cancer images. We distributed the survey to physicians working in Arkansas state hospitals. Using a scale of 1 to 7, ranging from authentic to fraudulent, we asked poll respondents to judge the veracity of each image. The discrimination and adversary losses while training are shown in Fig. 5 (Table 1).

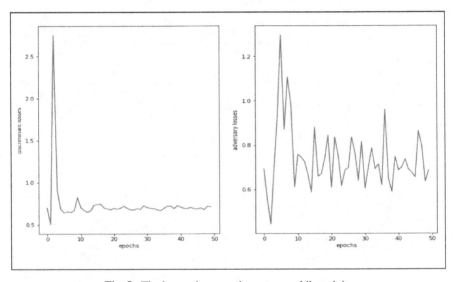

Fig. 5. The losses decrease the process while training.

Table 1. Some questions from the survey were asked of the participants.

Question	Option 1	Option 2
I am familiar with Deepfake	Yes	No
This image is real	Yes	No
This image has a breast tumor at the top	Yes	No
This image has a breast tumor in the bottom	Yes	No

4 Discussion

The capacity to discern between real and fake photographs is one benefit of employing GANs in this work. In the first part of the GANs network, the generators will generate breast tumor images, and in the second part, the discriminators will compare the rendered images with authentic images. The created photos will be sent back to the generator to be improved upon and the difference will be minimized if there is a significant disparity between the generated and real photographs. Science humans, such as physicians, are able to distinguish between real and fake breast tumor images. Therefore, we developed our hypothesis that physicians can identify fake images.

The performance of the discriminator affects the generator training's progress; that is why when the discriminator is optimal, the sent information to the generator won't be enough, and the training might fail due to vanishing gradients.

5 Conclusion

Today, machine learning ML techniques can detect cancer from an extensive set of medical images. The lack of medical images will affect the ML model's performance and prediction accuracy.

In this work, we utilized deep GANs to build a deepfake model to generate a fake breast cancer dataset to be used in ML techniques for cancer prediction. We tested the performance of our model by distributing a survey among some physicians to determine the scale of the ethnicity of some images. The results of using deepfake to generate the medical dataset showed that the images couldn't be differentiated from the original images.

Acknowledgments. We thank University of Arkansas/AHPCC center for providing a Pinnacle account which significantly improves the current work. National Science Foundation under Award No. OIA-1946391, DART SEED Research program.

References

1. Lambin, P., Rios-Velazquez, E., Leijenaar, R., et al.: Radiomics: extracting more information from medical images using advanced feature analysis. Eur. J. Cancer **48**(4), 441–446 (2012)
2. Sala, E., Mema, E., Himoto, Y., et al.: Unravelling tumour heterogeneity using next-generation imaging: radiomics, radiogenomics, and habitat imaging. Clin. Radiol. **72**(1), 3–10 (2017)
3. Kumar, V., Gu, Y., Basu, S., Berglund, A., Eschrich, S.A., Schabath, M.B., et al.: Radiomics: the process and the challenges. Magn. Reson. Imaging **30**(9), 1234–1248 (2012)
4. Mazurowski, M.A.: Radiogenomics: what it is and why it is important. J. Am. Coll. Radiol: JACR **12**(8), 862–866 (2015)
5. Kuo, M.D., Jamshidi, N.: Behind the numbers: decoding molecular phenotypes with radio-genomics–guiding principles and technical considerations. Radiology **270**(2), 320–325 (2014)
6. Bigos, K.L., Weinberger, D.R.: Imaging genetics–days of future past. Neuroimage **53**(3), 804–809 (2010)
7. Stoyanova, R., Pollack, A., Takhar, M., Lynne, C., Parra, N., Lam, L.L., et al.: Association of multiparametric MRI quantitative imaging features with prostate cancer gene expression in MRI- targeted prostate biopsies. Oncotarget **7**(33), 53362–53376 (2016)
8. Renard-Penna, R., et al.: Multiparametric magnetic resonance imaging predicts postoperative pathology but misses aggressive prostate cancers as assessed by cell cycle progression score. J. Urol. **194**(6), 1617–1623 (2015)

9. Mehta, S., Shelling, A., Muthukaruppan, A., Lasham, A., Blenkiron, C., Laking, G., et al.: Predictive and prognostic molecular markers for cancer medicine. Ther. Adv. Med. Oncol. **2**(2), 125–148 (2010)

10. Punnappurath, A., Brown, M.S.: Learning raw image reconstruction-aware deep image compressors. IEEE Trans. Pattern Anal. Mach. Intell. **42**, 1013–1019 (2019). https://doi.org/10.1109/TPAMI.2019.2903062

11. Cheng, Z., Sun, H., Takeuchi, M., Katto, J.: Energy compaction-based image compression using convolutional autoencoder. IEEE Trans. Multimedia **22**, 860–873 (2019). https://doi.org/10.1109/TMM.2019.2938345

12. Chorowski, J., Weiss, R.J., Bengio, S., Oord, A.V.D.: Unsupervised speech representation learning using wavenet autoencoders. IEEE/ACM Trans. Audio Speech Lang. Process. **27**(12), 2041–2053 (2019)

13. Karras, T., Laine, S., Aila, T.: A style-based generator architecture for generative adversarial networks. In: Proceedings of the IEEE Conference on Computer Vision and Pattern Recognition, pp. 4401–4410 (2019)

14. Karras, T., Laine, S., Aittala, M., Hellsten, J., Lehtinen, J., Aila, T.: Analyzing and improving the image quality of STYLEGAN. In: Proceedings of the IEEE/CVF Conference on Computer Vision and Pattern Recognition, pp. 8110–8119 (2020)

15. Zhang, X., Karaman, S., Chang, S.: Detecting and simulating artifacts in GAN fake images. In: 2019 IEEE International Workshop on Information Forensics and Security (WIFS), pp. 1–6. IEEE (2019)

16. Guarnera, L., Giudice, O., Battiato, S.: Deepfake detection by analyzing convolutional traces. In: Proceedings of the IEEE/CVF Conference on Computer Vision and Pattern Recognition Workshops, pp. 666–667 (2020)

17. Guarnera, L., Giudice, O., Battiato, S.: Fighting deepfake by exposing the convolutional traces on images. IEEE Access **8**, 165085–165098 (2020)

18. Zhu, J.Y., Park, T., Isola, P., Efros, A.A.: Unpaired image-to-image translation using cycle-consistent adversarial networks. arXiv preprint arXiv:1703.10593 (2017)

19. Masood, M., Nawaz, M., Malik, K.M., Javed, A., Irtaza, A., Malik, H.: Deepfakes generation and detection: State-of-the-art, open challenges, countermeasures, and way forward. Appl. Intell. (2022). https://doi.org/10.1007/s10489-022-03766-z

20. Shen, T., Liu, R., Bai, J., Li, Z.: Evaluation of an audio-video multimodal deepfake dataset using Unimodal and multimodal detectors. In: "Deep Fakes" Using Generative Adversarial Networks (GAN) (2018)

21. Khalid, H., Kim, M., Tariq, S., Woo, S.S.: Evaluation of an audio-video multimodal deepfake dataset using Unimodal and multimodal detectors. In: Proceedings of the 1st Workshop on Synthetic Multimedia - Audiovisual Deepfake Generation and Detection (2021). https://doi.org/10.1145/3476099.3484315

22. Guera, D., Delp, E.J.: Deepfake video detection using recurrent neural networks. In: 2018 15th IEEE International Conference on Advanced Video and Signal Based Surveillance (AVSS) (2018)

23. Pu, J., et al.: Deepfake Videos in the Wild: Analysis and Detection. arXiv. http://arxiv.org/abs/2103.04263 (2021)

24. Westerlund, M.: The emergence of deepfake technology: a review. Technol. Innov. Manag. Rev. **9**(11), 40–53 (2019). https://doi.org/10.22215/timreview/1282

25. Mangaokar, N., Pu, J., Bhattacharya, P., Reddy, C.K., Viswanath, B.: Jekyll: attacking medical image diagnostics using deep generative models. In: 2020 IEEE European Symposium on Security and Privacy (EuroS&P), Genoa, Italy, pp. 139–157 (2020). https://doi.org/10.1109/EuroSP48549.2020.00017

26. Thambawita, V., et al.: DeepFake electrocardiograms using generative adversarial networks are the beginning of the end for privacy issues in medicine. Sci. Rep. **11**(1), 21896 (2021). https://doi.org/10.1038/s41598-021-01295-2. PMID: 34753975; PMCID: PMC8578227
27. Swathi, P., Sk, S.: DeepFake creation and detection: a survey. In: 2021 Third International Conference on Inventive Research in Computing Applications (ICIRCA), Coimbatore, India, pp. 584–588 (2021). https://doi.org/10.1109/ICIRCA51532.2021.9544522
28. Goodfellow, I., et al.: Generative adversarial nets. In: Advances in Neural Information Processing Systems, pp. 2672–2680 (2014)
29. Radford, A., Metz, L., Chintala, S.: UnsuperFigure 13. Demonstration of results for transferring Ukiyo-e to photos. vised representation learning with deep convolutional genera- tive adversarial networks. arXiv preprint arXiv:1511.06434 (2015)
30. https://portal.gdc.cancer.gov/

A New Method for Microscopy Image Segmentation Using Multi-scale Line Detection

Fella Haddar[(✉)] and Djerou Leila

LESIA Laboratory, Mohamed Khider University, Biskra, Algeria
`fella.haddar@univ-biskra.dz`

Abstract. Image segmentation plays a crucial role in many biomedical imaging applications by automating and facilitating delineating of anatomical structures and different regions of interest. The objective of microscopic image segmentation is to accurately identify the boundaries of cells, cell nuclei, or histological structures in stained tissue images with various markers. Over the last few years, numerous techniques for segmenting microscopy images have been developed. Despite the advancements in segmentation techniques for microscopy images, there are still significant challenges when dealing with variations, high levels of noise, and variations in image features. (e.g., nucleus shape, cell size, concavity points between nuclei) and complexity of method parameter space. This paper introduces a new method for distinguishing nuclei and other biological structures in microscopy images. The proposed method exploits the multi scales line detection method, which presents an effective method for automatically extracting blood vessels from retinal images to detect boundaries of cells, cell nuclei and the concave points that split the contours into segments. The effectiveness of the method has been evaluated both qualitatively and quantitatively and compared with four other segmentation algorithms - the Chan & Vese model and three thresholding techniques, namely Otsu, Kapur, and Kittler method. These evaluations were conducted using a variety of publicly available datasets from the Broad-Bioimage Benchmark. The comparison results were analyzed using the Jaccard similarity index (JI) and the Dice coefficient (DSC) measure set agreement.

Keywords: Image segmentation · Microscopic images · Nucleus · Cell · Multi scales line detection

1 Introduction

Image segmentation is a fundamental task in image processing and computer vision to simplify the representation of an image, making it easier to analyze and process. Image segmentation has numerous applications, including object detection, tracking, recognition, and image compression. In other side microscopic images play an essential role in different fields, for example, biology (cell detection and tracking of cell lines) and medicine (decision-making for diagnosing diseases). The purpose of processing and analyzing microscopic images is to detect and segment cells/nuclei. However, the complex characteristics of this type of image make this operation more difficult because of

© The Author(s), under exclusive license to Springer Nature Switzerland AG 2024
A. Bennour et al. (Eds.): ISPR 2023, CCIS 1940, pp. 120–128, 2024.
https://doi.org/10.1007/978-3-031-46335-8_10

the low contrast between the background and the cells. The background is not uniform due to heterogeneous lighting, the presence and nature of noise and artifacts, the variability of the morphologies of the cells, aggregations and occlusions, the variability of the coloring, and the variability of illumination [3]. Each type of microscopy image has different image characteristics, so each type requires a specific detection and segmentation algorithm [4]. Most microscopic image segmentation methods are based on three popular strategies:

- Thresholding: several studies have used thresholding for microscopic image segmentation; thresholding is a straightforward and effective method for segmenting microscopic images.
- The contours active: many segmentation methods are based on active contours that minimize energy/cost function [2].
- Neural networks.

The purpose of contour detection is to identify the points in a digital image where the image's brightness changes abruptly. Edge detection of objects in the image is mainly concerned with extracting their edges from an image by identifying pixels with very high-intensity variation [4]. The primary forms of image intensity variations are typically classified as steps, lines, and junctions [5]. In recent years, research in the field of contour detection has been oriented toward the multiscale concept. On a small scale, responses originate many edges from noise. More edges that are interesting are those that also exist at larger scales. As the scale increases, most parasites are eliminated in the detected edges. On the other hand, the edges at larger scales are less localized than those at smaller scales. For good location and contour detection, a multiscale approach is required.

This paper presents a new segmentation method: the multiscale line detection method (MSLD). This segmentation method is used to detect blood vessels in retinal images because of its high quality and results. It has detected even small retinal vessels [1]. We adopted the MSLD method to segment the microscopy images. After an initial pretreatment step involving color deconvolution, the segmentation step consists of calculating the response of each pixel, which was the linear combination of responses of different scales, where in the end; we used a series of morphological operations. After using this method, we obtained very good results. We compared it with some segmentation algorithms (OTSU, CHENVESE, OTSU2D, and KITTLE). In addition, we validated the results using a set of metrics: for example, Dice and Jaccard. The rest of this paper is organized as follows: Sect. 2 presents the works used to segment microscopy images, Sect. 3 explains our process, and Sect. 4 discusses the results; finally, Sect. 5 concludes the paper with prospects for future work.

2 Related Work

The authors in [13] presented a new method of unsupervised segmentation of cell nuclei in histological tissues. The first step of this method is to perform an initial preliminary treatment. The second step is segmentation; this step consists of multi-level thresholding and a series of morphological operations. The required parameter for the method is the size of the minimum region, which is set according to the image's resolution. In [2], the authors developed a thresholding algorithm, the stable count threshold (SCT). The SCT algorithm provides much better segmentation than standard thresholding methods and, more importantly, is comparable to manual thresholding results.

Furthermore, In [4], the authors proposed an active mask framework that integrates various techniques, including regional, multi-scale, and multiresolution growth methods, as well as active contours. In [5], a fluorescence microscopy method for nuclei segmentation is proposed, which utilizes three-dimensional (3D) active contours with inhomogeneity correction. The main idea of the active contours based on the regions is to evolve a curve (2D) or a surface (3D) within the limit of an object by minimizing an energy function. [6] proposed a fully 3D kernel segmentation method using three-dimensional convolutional neural networks. Synthetic volumes with corresponding labeled volumes are automatically generated to form the network. There are many techniques to realize the multiscale. A technique proposed by Baris Sumengen and Manjunath [9] is based on a geometrical perspective. It is not necessary to estimate a scale locally. The primary goal of this technique is to identify edges that are present at both coarse and fine scales and to precisely locate them at the finest scale. The new method of variation segmentation is based on anisotropic diffusion. A vector field is employed in this anisotropic scattering scheme to locate the boundaries. This vector field is replaced by the field of multiscale Edge flow vectors to achieve multiscale image segmentation. Another approach to multiscale edge detection and segmentation is the Perona-Malik flow [10]. Anisotropic diffusion is a technique that aims to prevent smoothing around edges by using spatially adaptive variance, which is equivalent to applying Gaussian smoothing. In anisotropic diffusion, pixels located within homogenous regions are smoothed using a Gaussian filter with a large variance, while pixels near edges are smoothed at a smaller scale. A novel approach proposed by Tabb and Ahuja [11] involves designing a vector field for image segmentation and edge detection. The idea of designing a vector field for edge detection is similar to the Edge flow technique. To design the vector field for image segmentation and edge detection, Tabb and Ahuja [11] analyzed the neighborhood surrounding each pixel. The optimum neighborhood size changes from pixel to pixel for a multiscale representation and needs to be estimated. The technique proposed by Tabb and Ahuja [11] allows for the specification of a homogeneity parameter to determine the desired level of homogeneity within each region. Using this parameter, the neighborhood size and spatial scale are estimated adaptively for each pixel [12]. The proposed method utilizes elongated filters to measure the difference of oriented means of various lengths and orientations. Additionally, the technique uses a theoretical estimation of noise to reveal the significant edges of all lengths and orientations by employing a scale adaptive threshold and recursive decision process. The approach is capable of accurately localizing edges, even in images with low contrast and high levels of noise. The efficiency of this algorithm depends on a multiscale approach for computing all the "significantly different" oriented means present in an image.

3 Proposed Process

In this part, we will show the structure of our proposed process, who is defined as follow (Fig. 1):

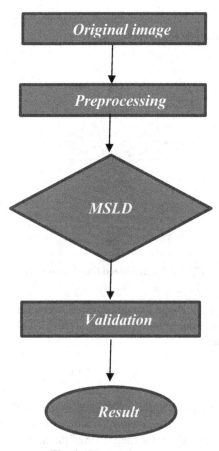

Fig. 1. Proposed process

3.1 Preprocessing

The pretreatment in this step consists of applying a set of operations, for example, noise reduction, contrast enhancement (based on the histogram), color correction, and image filtering (median, Gaussian). The main goal of image preprocessing is to improve the quality and accuracy of an image by removing artifacts, reducing noise, and enhancing features of interest.

3.2 Segmentation

This step applies the Multi-Line Scales Detection method to segment microscopy images. It is a retinal blood vessel segmentation method proposed by Ricci et al. and improved by Nguyen in 2013 [1]. The basic line detector is applied to the inverted green channel [1]. A window of size W * W is constructed for each image pixel, and the average gray level I_{avg}^{w}, is computed. Twelve lines of length W pixels oriented at 12 different directions passing through the centered pixels are identified; It then computes the average gray level of the pixels along each line. The line with the maximum value is called the winning line, and its value is defined as I_{max}^L [1] (Fig. 2).

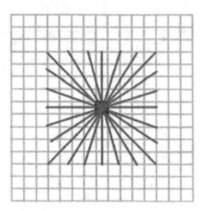

Fig. 2. A generalized line detector [1]

Nguyen et al. [1]. A generalized line detector has been proposed to enhance the line detection process by operating at multiple scales. The MSLD is based on varying the length of the aligned lines [1].

$$R_w^L = I_{max}^L - I_{avg}^w \tag{1}$$

where, $1 \leq L \leq W$

To improve the contrast of the response image at each scale, the values of the raw image are standardized to achieve a distribution with zero mean and unit standard deviation. This procedure is carried out for each scale of the image, as described in [8]. The standardization is defined as:

$$R\prime = \frac{R - R_{mean}}{R_{std}} \tag{2}$$

In [8], the response of each pixel is defined as a linear combination of the responses at different scales, defined as:

$$R_c = \frac{1}{n_L + 1}\left(\sum_L R_w^L + I_{igc}\right) \tag{3}$$

where, n_L is the number of scales and I_{igc} is the inverted green channel value corresponding to each pixel (Fig. 3).

3.3 Validation

This step uses a set of metrics to make the comparison between our segmentation method and the existing method. Among the existing metrics, we can cite Dice, which measures the similarity of two images and Jaccard, which compares the similarity and diversity (in) between images. Other metrics like distance from Hausdorff, is a widely used performance measure to calculate the distance between two point sets.

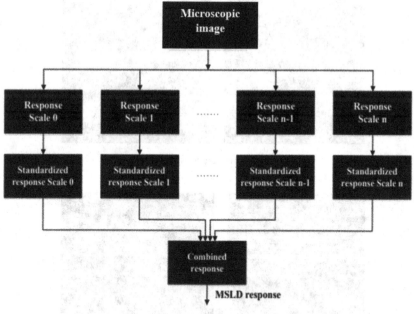

Fig. 3. MSLD process.

4 Evaluation and Result

This section presents the results obtained by the segmentation of microscopy images (MSLD). We used two types of images: histological and cytological. The collection currently contains 18 sets of cytological images, available at https://data.broadinstitute. org/bbbc/image_sets.html. To evaluate our method, we utilized images of synthetic cells comprising of five sets, each containing 20 images. These images consist of 300 objects, but due to varying probabilities of overlap and clustering, the objects appear differently in each of the five subsets. The images were generated with the SIMCEP simulating platform for fluorescent cell population images [14]. We used a collection of images (Fig. 6) for the histological image, available at http://mouse.brain-map.org [7] (Figs. 4 and 5).

To check the performance of the MSLD method, we used three classical segmentation methods. Otsu's method is used to perform automatic thresholding from the histogram shape of an image or to reduce a grayscale image to a binary image. Active contour segmentation using the Chan-Vese model is a widely used and flexible method that can effectively segment various types of images. The model has demonstrated its power in numerous applications in image processing and computer vision. Moreover, the Kittler method is a thresholding method used in digital image segmentation. It improves the segmentation process. In this work, we tested the performance on 20 images. Tables 1 and 2 show that the MSLD method is better than other methods of segmentation of the microscopy images according to the two similarity coefficients dice and Jaccard.

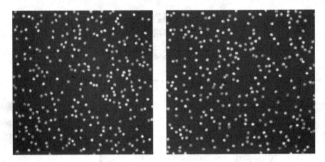

Fig. 4. Reference image

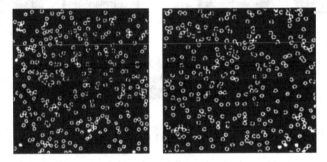

Fig. 5. MSLD result

Table 1. Comparison between the methods (for cytological image)

		MSLD	CHEN_VESE	OTSU	KITTLER
DICE	Average	0.8763	0.4380	0.8371	0.4025
	Median	0.8745	0.4403	0.8317	0.4033
	Std	0.0055	0.036	0.0165	0.0223
	Best	0.8895	0.4830	0.8652	0.4378
	Worst	0.8600	0.4135	0.8062	0.3746
JACCARD	Average	0.7783	0.2812	0.7216	0.2525
	Median	0.7770	0.2823	0.7119	0.2526
	Std	0.0053	0.0291	0.0211	0.0170
	Best	0.7885	0.3184	0.7624	0.2802
	Worst	0.7600	0.2606	0.6800	0.2305

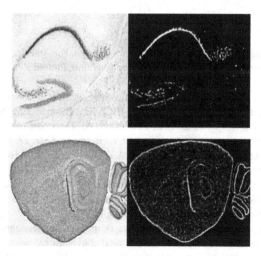

Fig. 6. Histological image and its result after the application of MSLD

Table 2. Comparison between the methods. (For histological image)

MSLD			CHEN_VESE	OTSU	KITTLER
DICE	Average	0.88	0.58	0.69	0.79
	Median	0.89	0.51	0.76	0.84
	Std	0.016	0.24	0.17	0.10
	Best	0.9	0.97	0.81	0.89
	Worst	0.86	0.32	0.39	0.65
JACCARD	Average	0.80	0.45	0.54	0.67
	Median	0.81	0.34	0.61	0.73
	Std	0.027	0.29	0.17	0.13
	Best	0.83	0.94	0.68	0.80
	Worst	0.76	0.29	0.25	0.49

5 Conclusion

This paper presented a new method of segmentation of microscopy images using Multi-Line Scales Detection (MSLD) where the results obtained are better than other methods (Chan & Vese model and three thresholding techniques: Otsu and Kittler method). The results were validated by the two metrics used for the evaluation: DICE and JACCARD. In future work, we will use an optimization method to improve the results and minimize the execution time.

References

1. Nguyen, U.T., Bhuiyan, A., Park, L.A., Ramamohanarao, K.: An effective retinal blood vessel segmentation method using multi-scale line detection. Pattern Recognit. **46**(3), 703–715 (2013)
2. Xing, F., Yang, L.: Robust nucleus/cell detection and segmentation in digital pathology and microscopy images: a comprehensive review. IEEE Rev. Biomed. Eng. **9**, 234–263 (2016)
3. Arteta, C., Lempitsky, V., Noble, J.A., Zisserman, A.: Detecting overlapping instances in microscopy images using extremal region tree. Med. Image Anal. **27**, 3–16 (2015)
4. Srinivasa, G., Fickus, M.C., Guo, Y., Linstedt, A.D., Kovacevic, J.: Active mask segmentation of fluorescence microscope images. IEEE Trans. Image Process. **18**(8), 1817–1829 (2009)
5. Lee, S., Salama, P., Dunn, K., Delp, E.: Segmentation of fluorescence microscopy images using three dimensional active contours with inhomogeneity correction. In: 2017 IEEE 14th International Symposium on Biomedical Imaging (2017)
6. Joon Ho, D., Fu, C., Salama, P., Dunn, K.W., Delp, E.J.: Nuclei segmentation of fluorescence microscopy images using three dimensional convolutional neural networks. In: IEEE Conference on Computer Vision and Pattern Recognition Workshops (CVPRW) (2017)
7. Allen Institute for Brain Science. Allen Reference Atlases. 2004–2006. http://mouse.brain-map.org
8. Bendaoudi, H., Cheriet, F., Pierre Langlois, J.M.: Memory Efficient Multi-Scale Line Detector Architecture for Retinal Blood Vessel Segmentation. Department of Computer & Software Engineering Polytechnique Montréal Montréal (Québec), Canada (9)
9. Sumengen, B., Manjunath, B.S.: Multi-scale edge detection and image segmentation. In: 13th European Signal Processing Conference 2005 (2005)
10. Perona, P., Malik, J.: Scale-space and edge detection using anisotropic diffusion. IEEE PAMI **12**, 629–639 (1990)
11. Tabb, M., Ahuja, N.: Multiscale image segmentation by integrated edge and region detection. IEEE Trans. Image Process. **6**(5), 642–655 (1997)
12. Galun, M., Basri, R., Brandt, A.: Multiscale Edge Detection and Fiber Enhancement Using Differences of Oriented Means. IEEE (2007)
13. Oliveira, D., et al.: Unsupervised segmentation method for cuboidal cell nuclei in histological prostate images based on minimum cross entropy. Expert Syst. Appl. **40**(18), 7331–7340 (2013)
14. Lehmussola, A., Ruusuvuori, P., Selinummi, J., Huttunen, H., Yli-Harja, O.: Computational framework for simulating fluorescence microscope images with cell populations. IEEE Trans. Med. Imaging **26**(7), 1010–1016 (2007)

Combination of Local Features and Deep Learning to Historical Manuscripts Dating

Merouane Boudraa and Akram Bennour[✉]

LAMIS Laboratory, Echahid Cheikh Larbi Tebessi University, Tebessa, Algeria
{merouane.boudraa,akram.bennour}@univ-tebessa.dz

Abstract. While paleographers face various challenges in the dating process of historical manuscripts, computer scientists encounter multiple obstacles in automating them. To address this problem, machine learning and deep learning techniques which have proven effective in other domains have been used. This study introduces a system that integrates two primary methods of feature extraction - deep learning and hand-crafted features. The Harris-detector was utilized to extract key-points from the manuscripts, and the K-means algorithm was applied to cluster them. From these clusters, patches of size nxn were extracted. Then, a densnet model was trained on these patches using transfer learning, finally we used majority voting on document patches to determine the date based on document level. The effectiveness of this approach was evaluated on two datasets, MPS and CLaMM, and achieved a CS of 96% and 65% and an MAE of 4.01 years and 20.1 years respectively for both datasets.

Keywords: Historical manuscripts · Dating · Deep learning · Machine learning · Feature extraction

1 Introduction

The dating of historical documents presents a crucial challenge for paleographers. In the past, gaining access to these manuscripts was difficult, but the advent of digital libraries has made thousands of old manuscripts available online.Unfortunately, many of these manuscripts are not properly dated, which reduces their usefulness. The complexity of the dating task requires a high level of expertise and the ability to discern features that are not easily visible to the human eye. To tackle this issue, computer science has developed various algorithms to automate the process. Although these algorithms have been effective in other domains, the task of dating historical manuscripts presents unique challenges that make it an interesting and exciting challenge for the computer science community.

The task of dating manuscripts can be viewed as a natural language processing (NLP) problem where the content of the document is processed to determine its date. In the past, optical character recognition (OCR) applications have been used to recognize the characters in the text, but due to the poor quality of historical documents, this method is not always accurate. Paleographers have the advantage of being able to use their

A. Bennour et al. (Eds.): ISPR 2023, CCIS 1940, pp. 129–143, 2024.
https://doi.org/10.1007/978-3-031-46335-8_11

knowledge of the historical context, which can be more useful in determining the date of a manuscript. Therefore, relying solely on an OCR application may not yield the best results.

More recently, the task has been considered a problem of computer vision and pattern recognition, where specific algorithms are used to classify documents based on the image itself. In recent years, pattern recognition and computer vision communities have developed many algorithms and methods that have been applied to this problem and have shown promising results. The image-based approach typically involves three steps: document image pre-processing, feature extraction, and classification. In the first step, the quality of the image is enhanced by removing noise and distortion, correcting for lighting and contrast, and separating text from the background. In the second step, specific features of the document image, such as the shape of the characters, the texture of the paper, and the layout of the text, are extracted and transformed into numerical representations. Finally, using the extracted features, the document is classified into a date range based on the patterns learned from a training set of dated documents. The accuracy of the classification depends on the quality of the features and the size and diversity of the training set.

In our paper, we aim to explore hand-crafted and deep-learning approaches for dating historical manuscripts using transfer learning. We begin by pre-processing the documents with various algorithms and then use both traditional and deep learning methods for feature extraction and classification. Our approach combines the expertise of paleographers and the power of computer vision and pattern recognition. We believe that our work can help improve the accuracy of dating historical manuscripts, making them more useful for historical research and preservation.

The paper is structured as follows: In Sect. 2, we discuss related work in the literature. Section 3 explains our methodology, including the dataset used for our experiments. Finally, in Sect. 4, we present our results, including a discussion and comparison with previous work.

2 Related Work

Dating historical documents is a challenging task that has been studied by researchers from various fields, including history, paleography, linguistics, and computer science. In recent years, there has been growing interest in applying deep learning and machine learning techniques to this problem, and several promising approaches have been proposed.

[1] provides a comprehensive overview of dating techniques for historical documents, including both traditional and computer-based methods, as well as an analysis of the datasets used in prior research.

In their research, He et al. [2] utilized a combination of hinge and fraglets features, as well as local support vector regression (SVR), to develop a model for accurately dating historical manuscripts. The findings of their study demonstrated a mean absolute error (MAE) of 35.4 and a cumulative score (CS) of approximately 93% for errors within a 75-year range.

He et al. [3] proposed another system for dating historical manuscripts, which employed histogram of orientations of handwritten strokes (HOHS) features to represent handwritten words' strokes. They trained a multi-label self-organizing map neural network on these features and evaluated the technique on the MPS dataset [2], achieving an MAE of 9.1 and a CS of 91.4%. Additionally.

In their work, He et al. [4] utilized a temporal pattern codebook (TPC) trained with self-organizing time map (SOTM) and combined it with PSD features to estimate the production dates of historical manuscripts from the MPS dataset [2]. They were able to achieve an MAE of 7.8 years and a CS of 93.3%. CLaMM dataset [10] was utilized in the 2017 International Conference on Document Analysis and Recognition (ICDAR) for competitions involving handwriting style and manuscript dating [11], there were four methods employed in the dating task, the T-DeepCNN System uses Deep Convolutional Neural Networks with residual learning [12] and batch normalization [13].

Table 1. Recent related work to historical manuscripts dating

Study	Features Used	Dataset	Evaluation Metrics	Results
He et al. [2]	Hinge and Fraglets features, local support vector regression	MPS	MAE, CS	MAE: 35.4; CS: 93% (for errors of < 75 years)
He et al. [3]	Histogram of Orientations of Handwritten Strokes (HOHS) features	MPS	MAE, CS	MAE: 9.1; CS: 91.4%
He et al. [4]	Temporal Pattern Codebook (TPC), PSD features	MPS	MAE, CS	MAE: 7.8 years; CS: 93.3% (for errors of < 25 years)
TDeepCNN System [11]	Deep Convolutional Neural Networks	CLaMM	Accuracy	59%
CK1 System [11]	CNN with two convolutional layers and two fully connected layers	CLaMM	Accuracy	54.15%
CK2 System [11]	CNN with three convolutional layers and two fully connected layers	CLaMM	Accuracy	54.85%
CK3 System [11]	VLAD descriptors	CLaMM	Accuracy	52.55%
P-CNN System [11]	Modified ResNet152 CNN	CLaMM	Accuracy	49.80%

The CK Systems are based on CNNs. The CK1 system uses a CNN with two convolutional layers and two fully connected layers, while CK2 uses a CNN with three convolutional layers and two fully connected layers. The CK3 system replaces the 1024-dimensional feature vector with a VLAD descriptor [14].

The P-CNN System uses a modified ResNet152 CNN to classify scripts. The network is trained on a dataset that is randomly partitioned into train and validation splits.

The ranking was based on the average accuracy (measured as the percentage of correctly classified images) evaluated on the complete test dataset, the winner was the T-Deep-CNN with 59% on dating task [11].

The results of different approaches for dating historical documents have shown that deep learning methods outperform the hand-crafted feature-based methods by a significant margin. Among the dating models evaluated on the MPS dataset [2].

According to research [1], Convolutional Neural Networks (ConvNets) that are pretrained with non-historical manuscript images and then adapted to the dating task have the best performance [1]. This was found in the study by [7].

The ConvNets adapted in this study were found to outperform Deep Learning (DL) nets trained solely on historical manuscript images. This could be attributed to the fact that the fine-tuned DL net has learned certain abilities that cannot be acquired from historical manuscript datasets alone. Furthermore, transfer learning capitalizes on the advantages of pre-trained nets, resulting in more effective, versatile, and superior features that are learned through fine-tuning with data from the task at hand.

Based on these findings, we utilized both the hand-crafted and deep learning approaches, along with transfer learning, to develop our system. A detailed explanation of our system is provided in the next section (Table 1).

3   Methodology

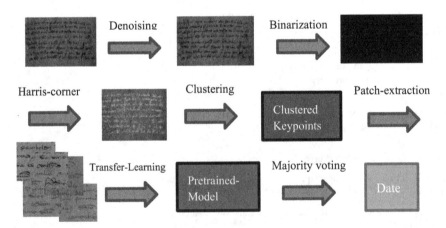

Fig. 1. Historical manuscript dating system design.

The community has previously conducted numerous experiments to analyze the development of handwritten styles. Hand-crafted features and deep learning are the two broad categories of feature extraction techniques. Our approach combines the strengths of both approaches. We utilized a Harris-corner detector to extract key-points from the manuscript and then clustered them using the K-means algorithm. We employed transfer learning to train a deep learning model on the patches extracted from the clustered key-points. Finally, a majority vote was performed on the patches to determine the date of the document. The combination of these two approaches aims to improve the accuracy of our method for dating historical manuscripts. Figure 1 illustrates the steps involved in our system design.

Before delving into the details of our method, which comprises three distinct steps: pre-processing, feature extraction, and classification, it is crucial to first take a closer look at the datasets utilized in our study.

3.1 Datasets

3.1.1 'Medieval Paleographical Scale (MPS) Dataset'

Our experiment made use of MPS dataset [2], which includes 3267 charters. Charters are an ancient type of document that were commonly used between 1300–1550 CE. This dataset is highly prevalent in academic literature [1, 3–6]. Table 2 provides an overview of how these charters are classified into 11 distinct categories.

Table 2. Number of samples in each period of the MPS dataset.

Period	1300	1325	1350	1375	1400	1425	1450	1475	1500	1525	1550
Samples	106	164	199	386	311	323	501	423	372	241	241

3.1.2 Classification of Medieval Handwritings in Latin Script (CLaMM)

The dataset we used contains annotated images of medieval handwriting from French catalogs [10]. It was utilized in the 2017 International Conference on Document Analysis and Recognition (ICDAR) for competitions involving tasks related to handwriting style and manuscript dating. As a result, the dataset is categorized based on these two tasks, with 12 classes and 15 classes respectively. The 15 classes in the manuscript dating category cover the period of 500–1600 AD. The training set consists of 6500 images, while the test set contains 3000 images [11, 15]. Table 3 illustrates the distribution of samples across different time periods.

3.2 Pre-processing

The pre-processing step plays a crucial role in eliminating the noise caused by poor image quality and ink traces on manuscript images (Fig. 2), as they can significantly

Table 3. Number of samples in each period of the CLaMM dataset.

Period	1000	1100	1200	1250	1300	1350	1400	1425	1450	1475	1500	1525	1550	1575	1600
Samples	671	78	417	125	192	233	184	169	226	675	262	258	80	39	31

Fig. 2. Original MPS and CLaMM manuscripts

reduce classification performance and increase errors. For this reason, it is essential to remove such ink traces from the images. To achieve this, we implemented Non-Local Means Denoising [8], By using this technique, we aimed to enhance the quality of the images and prepare them for feature extraction and classification in the subsequent steps.

3.2.1 Denoising

One of the denoising techniques we applied is Non-Local Means Denoising [8], which involves replacing a pixel's color with the average color of the surrounding pixels. By doing so, we can reduce the noise present in the image and make it clearer. We also utilized binarization techniques to convert the image into a binary format, which is easier to process for classification algorithms.

To demonstrate the effectiveness of the Non-Local Means Denoising technique, we compared the image before and after applying the denoising operation. The result is shown in Fig. 3, where it is evident that the noise present in the original image has been significantly reduced after applying the denoising operation. This cleaner image can now be used as input to the subsequent stages of the classification system, leading to improved accuracy and performance.

Fig. 3. Manuscripts after denoising

3.2.2 Binarization

Before feature extraction can be performed, the document images need to be binarized.

The Harris response image can be noisy and contain many small variations in response that are not true corners. By applying a threshold and binarizing the image, we can isolate the regions of high response that correspond to strong corners.

Binarization also simplifies further processing steps that may require binary input, such as extracting features or matching points across multiple images. By reducing the image to a binary representation, we can simplify and speed up these operations.

One approach that we used for this was the Canny edge detection method [9], which can effectively detect edges in the image and produce a binary representation of the document. The Canny edge detection algorithm works by identifying areas of the image where the gradient of intensity changes sharply, indicating the presence of an edge. This process helps to create a clear distinction between the text and background of the document, making it easier for feature extraction and classification algorithms to operate. We found that the Canny edge detection method produced good results, as shown in Fig. 4.

Fig. 4. Binarization using canny edge detection

3.3 Features-Extraction

Once the images have been pre-processed, the next step is to identify the regions of interest in each image. To accomplish this, we employ the Harris key-point detector [17] to extract informative junctions and corners, resulting in a set of regions that are relevant for handwriting analysis. These regions, or patches, are of size nxn, and by extracting features exclusively from these patches, we reduce the amount of data that needs to be processed. Additionally, the use of an interest point detector ensures that similar patches are sampled from different objects. The patches are subsequently clustered, resulting in a common invariant feature space across the entire dataset. Figure 5 displays a sample handwriting image with its extracted regions of interest. This process is repeated for all images under consideration, and the extracted patches are grouped using K-means clustering [16], a popular unsupervised machine learning algorithm. This algorithm is capable of clustering data into k number of clusters based on similarity, making it suitable for grouping similar images together based on features such as color, texture, or shape. We choose different k numbers of clusters on both datasets to cluster our extracted patches into k clusters.

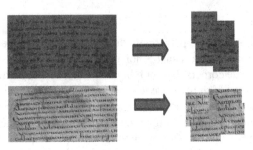

Fig. 5. Patches extraction

3.3.1 Harris-Corner

The Harris Corner Detection method [17] is based on the observation that corners or interest points in an image have distinctive and repeatable patterns of intensity changes in different directions. The method works by calculating the change in intensity in small neighborhoods around each pixel in the image, and then computing the second-moment matrix of the intensity gradients. The eigenvalues of this matrix provide a measure of the local image structure, and corners can be detected by identifying points where the eigenvalues are both large.

The Harris Corner Detection method has several advantages, including its simplicity and robustness to noise and image transformations. However, it can also be sensitive to changes in lighting conditions and may not work well for highly textured or cluttered images, Fig. 6 shows the keypoints Harris-corner detector highlighted key-points.

Fig. 6. Harris-corner detector on MPS and CLaMM documents.

3.3.2 Clustering (K-means)

After applying the Harris corner detection algorithm to an image, we get a set of corner points denoted as C, where each corner point $c \in C$ is represented as a 2D coordinate (x,y). We can then compute the centroids of these corner points using the following formula:

$$\text{centroid} = (1/|C|) * \Sigma c$$

where |C| is the number of corner points and Σc is the sum of all corner points. The resulting centroid is a 2D coordinate that represents the center of mass of the corner points.

After computing the centroids of the detected corners using Harris corner detection, these centroids can be used as input to the k-means clustering algorithm. The k-means algorithm [16] aims to partition a set of data points into k clusters, where each data point belongs to the cluster with the nearest mean or centroid. The algorithm iteratively updates the centroids of the clusters until convergence.

Let's denote the set of n data points by $X = \{x1, x2,..., xn\}$, and let $C = \{c1, c2,..., ck\}$ denote the set of k centroids. The goal of k-means is to find a partition of the data points into k clusters such that the sum of the squared distances between each data point and its assigned centroid is minimized. Mathematically, this can be expressed as:

$$\text{argmin}_C \sum_{i=1}^{k} \sum_{x \in S_i} |x - c_i|_2^2$$

where S_i denotes the set of data points assigned to cluster i, and $|x - c_i|_2$ is the Euclidean distance between data point x and centroid c_i.

By clustering the centroids (Fig. 7), we can group together similar corner points that may have been detected multiple times or that belong to the same object or region of interest in the image (Fig. 8). This can help reduce the number of keypoints and improve the accuracy of our dating system.

Fig. 7. Centroids calculated from Harris corners

Fig. 8. Clustered centroids

3.4 Classification

To train and classify our pre-processed dataset, we incorporated transfer learning by utilizing various pre-trained convolutional neural network (CNN) models. Due to the specific input size requirement of 224×224 for these models, we resized our patches of 550 to 224×224 before feeding them into the CNN models for training. This enabled us to leverage the capabilities of these pre-trained models while preserving the accuracy of our larger images. Among the models we used, Densnet produced the most favorable outcomes. Therefore, we have presented its architecture below.

3.4.1 Densnet

DenseNet-121 is a convolutional neural network architecture that was proposed in 2016 by [18] It is a variant of the deep residual network (ResNet) architecture that improves the flow of information and gradients through the network. DenseNet-121 has 121 layers in total and consists of dense blocks, transition layers, and a classification layer.

The dense blocks in DenseNet-121 consist of multiple layers, with each layer being connected to all previous layers in a feed-forward fashion. This dense connectivity allows for the reuse of features learned by previous layers, enabling more efficient training of the network. The transition layers in DenseNet-121 are used to reduce the dimensions of feature maps before passing them to the next dense block. This helps to reduce the number of parameters in the network and prevent overfitting.

Finally, the classification layer of DenseNet-121 consists of a global average pooling layer followed by a fully connected layer and a softmax activation function. The global average pooling layer aggregates the feature maps from the last dense block into a single vector, which is then fed into the fully connected layer for classification.

Overall, DenseNet-121 has demonstrated excellent performance on a variety of computer vision tasks, including image classification, object detection, and segmentation. Its dense connectivity and efficient use of parameters make it particularly suitable for training on limited data (Fig. 9).

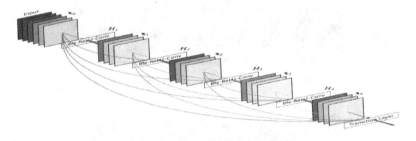

Fig. 9. Densenet architecture [18]

3.4.2 Majority Voting

Majority voting in image classification is a technique used to combine the predictions of multiple models or classifiers. In this approach, the final predicted class label for an image is determined by taking the most common prediction among predicted patches. Figure 10 explains this more.

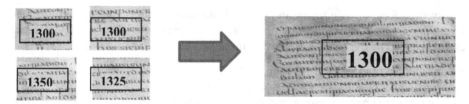

Fig. 10. Majority Vote Approach

4 Experiments and Results

In this experiment, we aim to determine the optimal values for the number of clusters (k), patch size, and pretrained model for our image dating system. We will test various combinations of these variables and evaluate their performance using metrics such as cumulative score and MAE on our MPS and CLaMM datasets.

First, we will test different patch sizes, such as 350×350, 550×550, and 750×750, to determine the optimal size for our image patches. We will then use these patches to test a variety of pretrained models, such as Densenet-121, ResNet-50, and Inception-v3, to determine which model produces the best dating results.

After selecting the optimal pretrained model and patch size, we will test various values of k clusters, such as 9, 16, and 25, to determine the optimal number of clusters for our dataset. We will evaluate the performance of these clusters using metrics such as cumulative score (1), MAE (2) and Accuracy (3).

To ensure the validity of our results, we will use a validation set of 20% of our datasets to test the different combinations of variables in each dataset. We will also repeat each experiment several times and calculate the average performance to reduce the impact of randomness.

Overall, the goal of this experiment is to identify the optimal combination of patch size, pretrained model, and number of clusters for image clustering, and to provide insights into how different variables impact the clustering performance. By doing so, we hope to improve our understanding of how to effectively the choice of these variables affects our system results.

$$\text{MAE} = 1/N * \sum |i = 1 \text{ to } N| \ |yi - ti| \tag{1}$$

The MAE measures the average magnitude of the errors between the estimated and actual production years, with lower values indicating better performance, where N is the total number of samples, yi is the estimated production year of the i-th historical manuscript, and ti is its actual production year. The absolute value of the difference between yi and ti is taken to ensure that the error is always positive.

$$\text{CS}(\alpha) = \text{Ne} \le \alpha/N \times 100\% \tag{2}$$

The cumulative score reflects the percentage of test images whose absolute error in estimating the production year is less than or equal to a threshold e. Therefore, the higher the CS, the better the performance of the dating model.

$$\text{Acc} = (1/n) * \Sigma i = 1 \text{ to } n(Ti/Ni) \tag{3}$$

Average accuracy is a metric used to evaluate the performance of a classification model in a multiclass classification problem. It is the average of the accuracy of each class, weighted by the number of samples in each class.

n is the number of classes
Ti is the number of correctly classified samples in class i
Ni is the total number of samples in class i.

4.1 Patch Size Experiment

In the patch size experiment, we tested three different patch sizes: 350×500, 550×550, and 750×750. The results showed that the patch size of 550×550 produced the best dating performance, with the highest cumulative score and the lowest MAE value as shown in Table 4. This suggests that using a patch size of 550×550 allows for capturing the relevant details in the images.

Table 4. Patch size experiment

Patch Size	Cumulative Score	MAE
350	88%	6 years
550	**96.6%**	**4.01 years**
750	90%	5.6 years

4.2 Pretrained Models Experiments

In the pretrained model experiment, we tested three different models: Densenet-121, ResNet-50, and InceptionResnet-v2 (Table 5). The results showed that Densenet-121 produced the best classification performance, with the highest cumulative score and the lowest MAE value. This suggests that Densenet-121 is better suited for image dating task, possibly due to its ability to capture more detailed and meaningful features from the images.

Table 5. Pretrained models experiments

Model	Cumulative Score (a = 25)	MAE
Resnet-50	93.1%	6.01 years
InceptionResnetV2	95.5%	5.3 years
Densnet-121	**96.6%**	**4.01 years**

4.3 Number of Clusters Experiment

In the number of clusters experiment, we tested three different values of k clusters: 9, 16, and 25. The results showed that using 16 clusters produced the best clustering performance, with the highest cumulative score and the lowest MAE value as shown in Table 6. This suggests that using a moderate number of clusters allows for capturing the relevant information in the images.

Table 6. Number of clusters experiments

Number of Clusters (k)	Cumulative Score (a = 25)	MAE
9	93.1%	5.2 years
16	**96.6%**	**4.01 years**
25	94%	5 years

4.4 CLaMM Dataset Experiment

For CLaMM dataset we pick 16 clusters and 550×550 patch size and try different models and the results are shown in Table 7.

Table 7. ClaMM dataset Experiments

Model	ACC
Resnet-50	55%
InceptionResnetV2	62%
Densnet-121	**65%**

Overall, the results of the experiments suggest that using a patch size of 550x550, Densenet-121 as the pretrained model, and 16 clusters produced the best clustering performance. These insights can help guide future historical manuscripts analysis tasks and improve the effectiveness of deep learning techniques in historical manuscripts classification.

4.5 Comparison and Discussion

We conducted a comparison between our system and the recent studies carried out on the same datasets as shown in Tables 8 and 9. We found that the proposed method is competitive with the state-of-the-art methods and could be a promising approach for this challenging task.

Table 8. Comparison of the proposed system with previous studies on CLaMM dataset using accuracy:

Method	Accuracy
T-DeepCNN [11]	59%
CK2 [11]	54.85%
CK3 [11]	52.55%
P-CNN [11]	49.80%
Proposed System	**65%**

Table 9. Comparison of the proposed system with previous studies on MPS dataset using MAE metric.

Method	CS	MAE
Fraglet and Hinge Features [2]	93.1%	35.4
Histogram of orientations (HOHS) [3]	91.4%	9.1
Temporal Pattern Codebook [4]	93.3%	7.80
Proposed System	**96.6%**	**4.01**

5   Conclusion

In conclusion, we have proposed a novel approach for dating historical manuscripts, which combines both traditional feature extraction techniques and modern deep learning methods to enhance the accuracy of the results. Our results suggest that the combination of hand-crafted features and deep learning can be a powerful approach for dating historical manuscripts. The use of region of interest extraction and clustering can help to reduce the amount of data and enhance the accuracy of the model. Additionally, transfer learning with a pre-trained model can improve the performance of the model with limited training data. However, there are still some limitations to our approach. The use of the Harris-corner detector may not be optimal for all types of manuscripts, and the clustering process may not always capture the most informative patches. Additionally, the pre-trained model used for transfer learning may not be optimal for all types of manuscripts or for all dating tasks. Future work can explore the use of other feature extraction techniques, clustering algorithms, and deep learning models to further improve the accuracy of our approach.

References

1. Omayio, E.O., Indu, S., Panda, J.: Historical manuscript dating: traditional and current trends. Multimed. Tools Appl. (2022). https://doi.org/10.1007/s11042-022-12927-8

2. He, S., Samara, P., Burgers, J., Schomaker, L.: Towards style-based dating of historical documents. In: International Conference of Frontiers in Handwriting Recognition (ICFHR), pp. 265–270 (2014)

3. He, S., Samara, P., Burgers, J., Schomaker, L.: A Multiple-Label guided clustering algorithm for historical document dating and localization. IEEE Trans. Image Process. **25**(11), 5252–5265 (2016). https://doi.org/10.1109/TIP.2016.2602078

4. He, S., Samara, P., Burgers, J., Schomaker, L.: Historical manuscript dating based on temporal pattern codebook. Comput. Vis. Image Underst. **152**, 167–175 (2016). https://doi.org/10.1016/j.cviu.2016.08.008

5. Hamid, A., Bibi, M., Siddiqi, I., Moetesum, M.: Historical manuscript dating using textural measures. In: International Conference on Frontiers of Information Technology (FIT), pp. 235–241 (2018). https://doi.org/10.1109/FIT.2018.00048

6. He, K., Zhang, X., Ren, S., Sun, J.: Deep residual learning for image recognition. In: Proceedings of the IEEE Conference on Computer Vision and Pattern Recognition, pp. 770–778 (2016)

7. Szegedy, C., Ioffe, S., Vanhoucke, V., Alemi, A.A.: Inception-v4, inception-resnet and the impact of residual connections on learning. In: Proceedings of the Thirty-First AAAI Conference on Artificial Intelligence, pp. 4278–4284 (2017)

8. Buades, A., Coll, B., Morel, J.M.: Non-local means denoising. Image Process. On Line **1**, 208–212 (2011)

9. Canny, J.: A computational approach to edge detection. IEEE Trans. Pattern Anal. Mach. Intell. **6**, 679–698 (1986)

10. Stutzmann, D.: Clustering of medieval scripts through computer image analysis: towards an evaluation protocol. Digit Medievalist J. **10** (2016). https://doi.org/10.16995/dm.61

11. Cloppet, F., Eglin, V., Helias-Baron, M., Kieu, V.C., Stutzmann, D., Vincent, N.: ICDAR 2017 competition on the classification of medieval handwritings in latin script. In: 14th IAPR International Conference on Document Analysis and Recognition (ICDAR) 2017, Kyoto, pp. 1371–1376 (2017). https://doi.org/10.1109/ICDAR.2017.224

12. He, K., Zhang, X., Ren, S., Sun, J.: Deep Residual Learning for Image Recognition, arXiv151203385 Cs (2015)

13. Ioffe, S., Szegedy, C.: Batch Normalization: Accelerating Deep Network Training by Reducing Internal Covariate Shift, arXiv150203167 Cs (2015)

14. Christlein, V., Bernecker, D., Angelopoulou, E.: Writer identification using VLAD encoded contour-Zernike moments. In: 2015 13th International Conference on Document Analysis and Recognition (ICDAR), Nancy, pp. 906–910 (2015)

15. Cloppet, F., Eglin, V., Kieu, V.C., Stutzmann, D., Vincent, N.: ICFHR2016 Competition on the classification of medieval handwritings in latin script. In: Proceedings of the International Conference on Frontiers in Handwriting Recognition (ICFHR) (2016)

16. Jin, X., Han, J.: K-Means Clustering. In: Sammut, C., Webb, G.I. (eds.) Encyclopedia of Machine Learning. Springer, Boston (2011). https://doi.org/10.1007/978-0-387-30164-8_425

17. Harris, C., Stephens, M.: A combined corner and edge detector. In: Proceedings of the 4th Alvey Vision Conference, pp. 147–151 (1988)

18. Huang, G., Liu, Z., Van Der Maaten, L., Weinberger, K.Q.: Densely connected convolutional networks. In: Proceedings of the IEEE Conference on Computer Vision and Pattern Recognition, pp. 4700–4708 (2017)

19. Wu, K., Otoo, E., Suzuki, K.: Two strategies to speed up connected component labeling algorithms. IEEE Trans. Image Process. **9**(11)

Data Mining

Toward an Ontology of Pattern Mining over Data Streams

Dame Samb[1(✉)], Yahya Slimani[2], and Samba Ndiaye[3]

[1] Management Department, UIDT, Cite Malick SY, N2, BP: A967, Thies, Senegal
`dsamb@univ-thies.sn`
[2] ISAMM, Tunis, Tunisia
[3] Mathematics and Computer Science Department, Cheikh Anta Diop University,
Dakar, Senegal
`samba.ndiaye@ucad.edu.sn`

Abstract. Pattern Mining over Data Stream (PMDS) is part of the most significant task in data mining. A major challenge is to define a representational framework that unifies PMDS algorithms dealing with different pattern types (frequent itemset, high-utility itemset, uncertain frequent itemset), using different methods (test-and-generate, pattern-growth, hybrid) and different window models (landmark, sliding, decay, tilted) in a uniform fashion. This will help standardize the process and create a better understanding of the algorithm design, provide a base for unification and research opportunities. It also facilitates the variability management and allows the derivation of tools for wide experimentation. In this publication, we propose a reference ontology to formalize the domain knowledge around PMDS. The design process of the ontology followed leading practices in ontology engineering. It is aligned to the most popular data mining and machine learning ontologies and thus, represents a major contribution toward PMDS domain ontologies.

Keywords: Data stream · Pattern mining · Ontology of data mining · Domain ontology · Knowledge representation

1 Introduction

In the past few years, there has been an increase in production of data in the form of stream causing the emergence of new applications. Data stream is inherently dynamic. It is continually updated and arriving at high speed in a disorderly manner. On the one hand, these aforementioned characteristics make patterns mining extremely difficult, due to the impossibility to maintain all the data in memory. The research in this new domain is classified as Pattern Mining over Data Stream (PMDS).

Pattern mining in data stream presents many challenges. The need to consider different window models (landmark, sliding, decay, etc.) and different pattern types (high-utility, frequent, uncertain frequent, etc.) makes the task even more complex. Over the last twenty years, extensive research has been done producing well-known algorithms [2, 10, 11, 14, 18, 22–25, 34]. However, further progress in the field is hindered by the lack of a standard framework that is widely accepted in the domain of PMDS. Then, constructing an ontology that provides formal standardized definitions for the principal PMDS entities would be of benefit to the whole research community.

Several ontologies have been proposed in the last decade as a representation and model in the domain of machine learning and data mining. *Onto-DM* (an Ontology of Data Mining) defines the most essential data mining entities in a three-layered ontological structure [26]. *DMOP* (Data Mining OPtimization ontology) has been developed to support meta-mining (meta-analysis) of complete data mining experiments in order to extract workflow patterns [16, 20]. *Exposé* has been designed to describe machine learning experiments in a standardized fashion [32]. Finally, the *ML-Schema* is a top-level ontology for representing and interchanging information on machine learning experiments in a concise, unambiguous and computationally processable manner [29]. The mentioned ontologies are general-purpose and include the definition of the main data mining tasks, data and algorithms. None of them deeply cover the area pattern mining and support data in the form of stream. Also, all of the approaches are superficial. The internal logic of the algorithm is not well describe, constitute a black box of only inputs and outputs. To take up this challenge, we propose in this paper a reference ontology that expresses, organizes and represents the domain knowledge of PMDS. The main contributions of our proposal are:

- A representation of the stream entity classes (stream, windows, model specification, data type, data stream specification);
- A representation of the algorithm entity classes (specification, implementation, execution, accuracy, approach, update type);
- A representation of the key PMDS entities (stream, windows, tasks, patterns, algorithms) in the context of a general framework.

The rest of the paper is organized as follows: Sect. 2 presents the PMDS ontology design process. Section 3 evaluates the proposed ontology. Section 4 is the conclusion of the paper and provides some potential research perspectives.

2 The PMDS Ontology

This part of the paper covers the PMDS ontology motivations and design principles. Next, it describes the ontology core concept with particular attention to the stream concept. Finally, it suggests the structure of the key classes of the PMDS ontology and discuss the alignment to related ontologies.

2.1 The PMDS Ontology: Motivations and Design Principles

There are many reasons that motivate the development an ontology in the PMDS domain. Firstly, the area is expanding and provides a major challenge in developing a general framework for mining patterns of different types, methods and window models. The ontology would formalize the basic entities (e.g., data stream, algorithm, task) and define the relations between the entities. Secondly, the majority of proposed data mining ontologies are general-purpose, aimed at covering the fundamental data mining tasks (probability distribution estimation, predictive modelling, clustering, pattern mining) and are superficial on describing the algorithms. Finally, ontology modelling is of great interest to software product line engineering [6,13]. The development of a PMDS ontology can ease the process for recovering a reference architecture from which the architecture of each product line can be easily and systematically derived for wide experimentation.

The PMDS ontology design process follows leading practices in the data mining and machine learning state-of-the-art ontologies [7,19,26,29,32], such as the Open Biomedical Ontologies (OBO) Foundry principles, which are widely accepted in the biomedical domain [30]. It uses the Basic Formal Ontology (BFO) [3] as the upper-level ontology, and the relations are reused from the Relations Ontology (RO) [31]. Furthermore, the PMDS ontology reuses appropriate classes from other ontologies that act as mid-level ontologies including the Information Artefact Ontology (IAO) [9], the Ontology of Biomedical Investigations (OBI) [4] and the OntoDT ontology for the representation of datatypes [27]. Table 1 establishes the specifications from which the ontology was derived through the following four aspects: the domain including competency questions, the purpose, the users and the sources of knowledge.

2.2   Structure and Implementation

This section provides a controlled vocabulary of basic concepts for the description of the PMDS domain. Indeed, it is useful to introduce the terms that will be found in formalisms modeling the domain knowledge, establishing what should be comprehend about those terms. To this end, from the competency questions detailed in Table 1, we can deduct the PMDS ontology basic entities such as *Pattern, Task, Data Stream, Window Model, Algorithm*. Enumerating, the terms related to PMDS, the classes and their relationship were defined inside the Protégé[1] tool using the code Ontology Web Language (OWL). Protégé is an open-source ontology editor that supplies: i) a graphical visualization tool (OntoGraf); ii) an automated reasoner (Pellet) to detect inconsistencies; and, iii) an interface for querying an searching an ontology (DL Query and SPARQL Query).

Data Stream and Window Models. A *data stream* $S =< T_1, T_2, \ldots, T_i, \ldots >$ is a sequence of incoming transactions ordered according to their arrival time. A window is a subsequence between the i-th and j-th arrived transactions, denoted by $W[i,j] =< T_i, T_{i+1}, \ldots, T_j >$, $i \leq j$. Several window models exist in the literature and the best known are [21]: *landmark, sliding, damped, tilted*.

---

[1] https://protege.stanford.edu/.

Table 1. Specifications of the PMDS ontology.

Competency question	1. Does the ontology enable to represent the PMDS domain knowledge from the state-of-the-art?
	2. What are the tasks (FIM, HUIM, UFIM) and patterns related to PMDS algorithms?
	3. What are the most used PMDS algorithms and their approach?
	4. What are the most used PMDS algorithms and their accuracy?
	5. What are the most used PMDS algorithms and their window model?
	6. What are characteristics of the most used PMDS algorithms?
	7. What are the characteristics (or typology) of a given PMDS algorithms?
	8. List of PMDS algorithms satisfying given characteristics?
Purpose	Describe the basic entities of PMDS
	Provide a basis for unification and standardization;
	Formalize the knowledge about the domain;
	Ease the understanding and design of algorithms
Users	Researchers (beginner and expert) concerned by the subject (PMDS)
Sources of knowledge	We have collected through Google Scholar and other scientific databases (ScienceDirect, SpringerLink, IEEE Xplore, Citex, etc.) about 410 papers published between 2002 and 2021 (up to December 31st) in prestigious journals, workshops or international conference proceedings on pattern mining in data stream. The following search string was used: ("data stream" AND "pattern mining") OR ("data stream" AND "frequent itemset mining") OR "data stream mining" OR "high utility data streams" OR" high average utility data streams" OR "uncertain data stream". We follow a paper selection process in three steps: Selection by inclusion and exclusion criteria, Selection by citation score, Selection by full reading. Finally, 65 papers (corresponding to 68 algorithms) were selected covering this different pattern types: frequent, closed and maximal itemset in the support framework; HUI and HAUI in the utility framework; PFI and EFI in the uncertain framework

Landmark Window: In this model $W[s,t]$, transactions are considered from a starting point s, called landmark (special case is when $s = 1$), to the current point t.

Sliding Window: In this model, transactions are considered in a window $W[t - w+1, t]$, where w is the window size from the current time point t. As time goes, each window moves and transactions arriving before the time point $t - w + 1$ are discarded.

Damped Window: It associates to each transaction a weight which decreases over time. When a new transaction arrives, the support count of the previous mined patterns is multiplied by a *decay factor* d, $0 < d < 1$ to reduce their contribution.

Tilted Window: In this model, transactions are considered in varying size windows. Each window corresponds to different granularity. In the simplest version, each window size is twice of that more recent neighbor.

The Fig. 1 shows the basic classes of the stream ontology and their relations (in our commentary, the class names are written in italic and the relation names are in small capital). The main elements are the stream, the data stream specification and the (window) model specification. The *stream* and *window* classes are defined as a subclass of the *information content entity* class provided by the IAO ontology [9]. A *stream* HAS-PART a *window* and is part of a *stream*. The *stream specification* IS-ABOUT the *stream* and HAS-PART the *sequence datatype* from the ontoDT class [27]. The *sequence datatype* class refer to the datatypes of the *data examples* in a *window*. The *data specification* specifies the (transaction) data used for the PMDS algorithm. The *model specification* IS-ABOUT the *window* entity. We distinguish between *the landmark*, the *decay*, *titled* and the *sliding window* (e.g., fixed-size, variable-size) sub-classes.

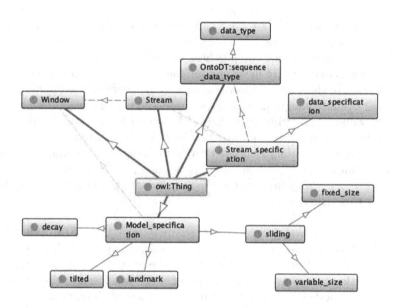

Fig. 1. The basic stream classes and their relations.

Tasks and Patterns. Tasks aim to produce some type of patterns from data stream. A pattern is straight associated to the output of a task or an PMDS algorithm. Conforming to the pattern produced, we will focus on three main tasks: Frequent Itemset Mining (FIM), High-Utility Itemset Mining (HUIM) and Frequent Itemset Mining from Uncertain data (UFIM).

Frequent Itemset Mining (FIM): The problem was proposed in the early nineties in *Basket Market Analysis* [1], where the database records *items* bought by a consumer as a *transaction*. Given minimum support threshold, a *Frequent Itemset* (FI) is a group of items contained in a significant number of transactions. *Apriori* [1], the first heuristic to find frequent itemsets followed *test-and-generate* technique where, in each level, candidate itemsets were formed from the already mined (frequent) itemsets. To address the main limitations of this heuristic, *FP-Growth* [15] the first *pattern-growth* algorithm was proposed. Its main improvements are keeping track the frequent itemsets without candidate generation phase and storing compressed information about patterns in a tree structure called *FP-tree* [15]. In contrast, *hybrid* algorithms such as *Eclat* [35] use a levelwise depth-first search strategy and a *vertical database representation* called *TidList* [35]. *Closed Itemset* (CI [28]) and *Maximal Itemset* (MFI [5]) are *concise representations* of frequent itemsets which decrease the huge number of FIs computed by the above-mentionned algorithms.

High-Utility Itemset Mining (HUIM): In FIM, an itemset at most appears onetime in a transaction and all items are treated with the same importance/weight/price. However, in a supermarket, items have different importance/price and one customer can purchase multiple quantities of an item. Thus, the framework is extended to develop a mining model [33] discovering the *High-Utility Itemset.* In the utility framework, each item is defined the *internal utility,* i.e., number of units bought in each transaction and the *external utility,* i.e., the unit profit value. High-Utiliy Itemsets (HUI) are all itemsets that have a *utility* higher than a given threshold in a database (i.e., itemsets generating a high profit). *High Average-Utility Itemsets* (HAUIs) mining is an emerging technique and a variation of the HUIM problem. It is based on an alternative measure called the *average-utility* (au) that is used to evaluate the utility of itemsets by considering their lengths [17].

Frequent Itemset Mining from Uncertain data (UFIM): In FIM and HUIM, patterns are discovered from data in which, users definitely know whether an item is present (or absent from) in a transaction in the database. However, in several real-life situations the presence or absence of items is uncertain since collected data is often imperfect, inaccurate, or may be collected through noisy sensors. Existing works on frequent itemset mining over uncertain data fall into two groups: *Expected-support Frequent Itemsets* (EFI [12]) and *Probabilistic Frequent Itemsets* (PFI [8]). In addition, it is possible to distinguish between the *tuple uncertainty model* and *attribute uncertainty model.* The former considers that each tuple or transaction is associated with an *existential probability* (a value in [0,1]), which indicates the chance that the transaction exists in the database. The latter attaches the *existential probability* to each attribute or item appearing in a transaction.

Algorithms and Methods. An (PMDS) algorithm transforms a task description into actions and produces some type of patterns from a data stream. Different methods are employed to achieve this goal. The *method* is the approach (technique) used to discover the patterns. The (*Apriori, pattern growth* and *hybrid*) methods consistently accepted in traditional pattern mining algorithms extends to PMDS algorithms.

Furthermore, algorithms can be characterized by the number of transactions ($N \geq 1$) handled in each update operation. This leads to treat a single transaction (*Per transaction update*) or a batch of N transactions (*Per batch update*). A more fine charaterization (for batch update) is to consider the update frequency in terms of number (e.g., every 1000 transactions) or in terms of time (e.g., weekly, monthly). The algorithms of the first group (respectively those of the last group) are referred as *Transaction-Sensitive* (respectively *Time-Sensitive*). In the sliding model, we have two variants: *Fixed-size Windows* (FsW) and *Variable-size windows* (Vsw).

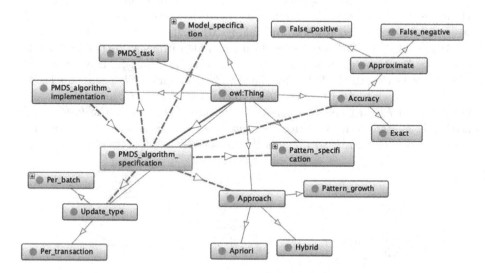

Fig. 2. The basic classes of the algorithm entity and their relations.

The Fig. 2 shows the basic classes of the algorithm entity and their relations. Here, we consider the specification aspect of the PMDS algorithm. A PMDS algorithm is represented as a subclass of the IAO plan specification. The *algorithm specification* class is connected to the *pattern specification, model specification, approach* and *PMDS task* classes via the relation HAS-PART. The *algorithm implementation* IS-A-CONCRETIZATION of an *algorithm specification*, implemented as a computer program, and written in a programming language. The algorithms may have various characteristics. One of these is the *accuracy* connected to the algorithm class via the relation HAS-PART. The accuracy class has two subclasses for representation of *exact* and *approximate* algorithms. Approximate algorithms may produce *false-positive* error. They may also produce *false-negative* or both types of error. Approximate algorithms accuracy depends on the error parameter value. Another characteristic of a PMDS algorithm is the update type. The *update type* is associated to the algorithm class via the relation HAS-PART. We distinguish between *the per transaction* and the *per batch* (e.g., *Time-Sensitive, Transaction-Sensitive*) update sub-classes.

2.3   The Structure and Alignment with Related Ontologies

Figure 3 presents the structure of the PMDS ontology with the key classes and most important relations. We consider three aspects of a PMDS algorithm: the *specification,* the *implementation,* and *the execution.* A PMDS *algorithm speci-fication* HAS-PART a *PMDS task* represented as an IAO objective specification subclass. Next, we have the *algorithm specification* directly related to the *update type, model specification, pattern specification, accuracy* and *approach* classes via the HAS-PART relation. An *algorithm execution* is a step-by-step function that REALIZES the *algorithm implementation* related to an *algorithm specification* via the IS-CONCRETIZATION-OF relation. An *algorithm execution* HAS-INPUT a stream, HAS-OUTPUT a *pattern,* and finally ACHIEVES-PLANNED-OBJECTIVE of a *PMDS task.* The PMDS tasks, patterns, and algorithms are dependent of the datatype. A *pattern specification* HAS-PART a *sequence data type.* It is subclassed at the first level with the following classes: *HUI, FI* and *UFI.* These subclasses are further subclassed with *HAUI* class in the case of *HUI, CI* and *MFI* classes in the case of *FI, EFI* and *PFI* classes in the case of UFI. Next, the *PMDS task* is directly related to the *sequence data type* class via the HAS-PART relation and is sub-classed with *FIM, UFIM* and *HUIM* classes.

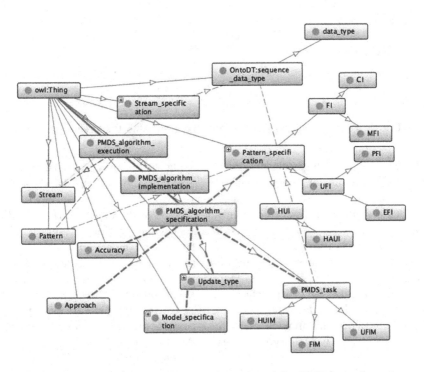

Fig. 3. The structure of the key classes of the PMDS ontology.

Table 2 shows the matching of the terms present in the PMDS ontology (OntoPMDS) with the current ML/DM ontologies [26,29]. It highlights how ontoPMDS is consistent and aligned with above-mentioned ontologies such as DMOP, OntoDM, ML-Schema and Exposé. Entities described in these ontologies can be described fully using the proposed ontology which in contrast, extends the domain with concepts particular to PMDS.

Table 2. Mapping between the core terms of PMDS ontology and their related ontologies.

ML-Schema	OntoDM	DMOP	Exposé	OntoPMDS
Task	Data mining task	DM-Task	Task	PMDS Task
Algorithm	Data mining algorithm	DM-Algorithm	Algorithm specification	PMDS algorithm
Implementation	Data mining implementation	DM-operator	Algorithm implementation	PMDS algorithm implementation
Run	Data mining execution	DM-operation	Algorithm execution	PMDS algorithm execution
Data	Data item	DM-data	Data	Data
Data characteristic	Data specification	Data characteristic	Dataset specification	Data specification
Dataset	Dataset	Dataset	Data	Window
Dataset characteristic	Dataset specification	Dataset characteristic	Data quality	Stream specification
Model	Generalization	DM-Model	Model	Pattern
Does not Apply	Does not Apply	Does not Apply	Does not Apply	Stream
Does not Apply	Does not Apply	Does not Apply	Does not Apply	Model specification
Does not Apply	Does not Apply	Does not Apply	Does not Apply	Update type
Does not Apply	Does not Apply	Does not Apply	Does not Apply	Accuracy

3   Evaluation

The PMDS ontology was evaluated from four different aspects. First, we use the ontology metrics data from the Protégé software. Then, the syntax of our ontology is tested using the Pellet reasoner. After that, we assess the ontology in regard to the list of competency questions using SPARQL queries. Finally, we demonstrate the utility of the PMDS ontology by a use case, i.e., its knowledge graph application for analyzing algorithms.

Figure 4 presents the ontology statistical metrics. The PMDS ontology contains 44 classes among which seven (Accuracy, Approach, Model specification, Pattern specification, PMDS algorithm specification, Tasks, update type) match to the core concepts identified in our competency questions (see Table 1). It was populated by 68 instances representing the most used PMDS algorithms. This set (ontology classes with the set of instances) constitutes the knowledge base, which further serves as input for the knowledge graph.

We use the automated reasoner Pellet to determine whether the ontology has any potential inconsistencies. All the classes were parsed and the output shows that they are syntactically correct (see Fig. 5).

The SPARQL query language (within the Protégé user interface) was used to answer the different competency questions. For example, Fig. 6 shows the responses to competency questions #1. The answer given consists of the list of the ontology classes (CI, Apriori, PMDS algorithm specification, decay, PFI, etc.).

We select Neo4j[2] (Neo4j Desktop Version 1.5.2), the most widely used open-source Graph Database Management System (GDBMS) as the knowledge graph

---

[2] https://neo4j.com/download/.

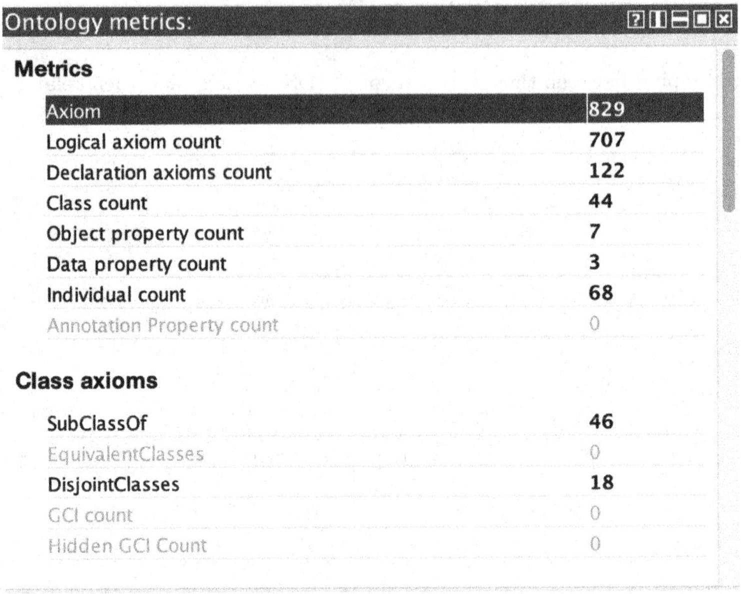

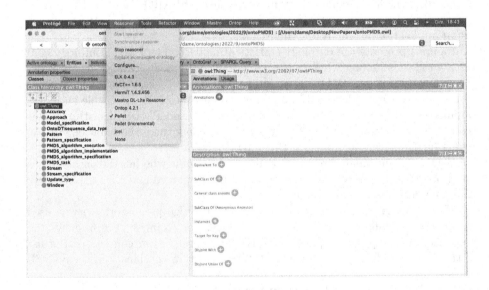

Fig. 4. Statistical metrics (screenshot from Protégé) of the PMDS ontology.

Fig. 5. The ontology classes syntactically verified using the Pellet reasoner.

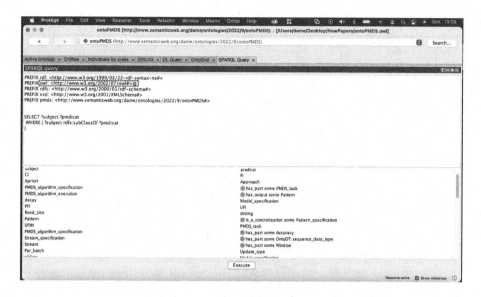

Fig. 6. SPARQL query answer to the competency questions.

analytic and visualization tool. There are two main steps to transfer data from Protégé to Neo4j. The first step is to export the ontology and instances from Protégé to RDF/XML file. The second step is to import the RDF/XML file into Neo4j through a plugin called neosemantics[3] (n10s).

Based on data from Neo4j, 883 triples were imported and parsed. Additionally, as shown in Fig. 7 in the knowledge graph below, 112 resource nodes and 479 relations were created. The graph shows the instance nodes colored in orange and their interrelations with the class nodes in green.

The Neo4j query language known as CQL (Cypher Query Language) is similar to SQL and lets to retrieve data from the graph. Figure 8 shows the output of a query which list the most cited (top ten) PMDS algorithms in the knowledge graph.

---

[3] https://neo4j.com/labs/neosemantics/.

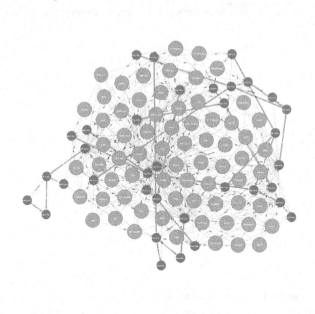

Fig. 7. The knowledge graph generated in Neo4j.

neo4j$ match(n:NamedIndividual) with n.hasName as Algorithm, n.hasNumberofCitation as Citation, n.PublicationY...

Algorithm	Citation	year	Description
"Lossy Counting"	1863	2002	["NamedIndividual", "PMDS_algorithm_specification", "landmark", "FI", "FIM", "Apriori", "False_positive"]
"FP-Stream"	773	2003	["NamedIndividual", "PMDS_algorithm_specification", "Pattern_growth", "FI", "FIM", "False_positive", "tilted"]
"IHUP"	818	2009	["NamedIndividual", "Exact", "PMDS_algorithm_specification", "Pattern_growth", "fixed_size", "HUIM", "Transaction_sensitive", "HUI"]
"estDec"	458	2003	["NamedIndividual", "PMDS_algorithm_specification", "FI", "FIM", "Apriori", "False_positive", "decay"]
"Moment"	409	2004	["NamedIndividual", "Exact", "PMDS_algorithm_specification", "Pattern_growth", "FIM", "fixed_size", "Transaction_sensitive", "CI"]
"MFI-TransSW"	227	2009	["Hybrid", "NamedIndividual", "Exact", "PMDS_algorithm_specification", "FI", "FIM", "fixed_size", "Transaction_sensitive"]
"MFI-TimeSW"	227	2009	["Hybrid", "NamedIndividual", "Exact", "PMDS_algorithm_specification", "FI", "FIM", "fixed_size", "Time_sensitive"]
"FDPM"	222	2004	["NamedIndividual", "PMDS_algorithm_specification", "landmark", "FI", "FIM", "Apriori", "False_negative"]

Fig. 8. Top ten PMDS algorithms.

4 Conclusion

Through this paper, we propose a reference ontology which expresses, organizes and represents the domain knowledge of the PMDS. Entities represented in the PMDS ontology include data stream (windows, stream specification, sequence data type), PMDS tasks (FIM, HUIM, UFIM), PMDS algorithms and their components (Approach, accuracy, model specification, update type), and patterns (FI, HUI, UFI). The design process follows the key steps found in the datamining and machine learning updating ontologies such as OntoDM, OntoDT, ML-schema, DMOP, etc. The ontology has been intensively evaluated considering the practices in ontology engineering and is revealed fully aligned with the other related domain ontologies. Currently, we are working on the further development of different aspects of the PMDS ontology. Some of these will require a holistic view that encapsulates the three different tasks and patterns in a unified scheme. Finally, investigating the ontological aspects in modelling the variability (in the domain of Software Product Line Engineering) of PMDS algorithms, is one our long term research objective.

References

1. Agrawal, R., Imielinski, T., Swami, A.: Database mining: a performance perspective. IEEE Trans. Knowl. Data Eng. **5**(6), 914–925 (1993)
2. Ahmed, C.F., Tanbeer, S.K., Jeong, B.S., Lee, Y.K.: Efficient tree structures for high utility pattern mining in incremental databases. IEEE Trans. Knowl. Data Eng. **21**(12), 1708–1721 (2009)
3. Arp, R., Smith, B., Spear, A.D.: Building Ontologies with Basic Formal Ontology. MIT Press, Cambridge (2015)
4. Bandrowski, A., et al.: The ontology for biomedical investigations. PLoS ONE **11**(4), e0154556 (2016)
5. Bayardo, R.J., Jr.: Efficiently mining long patterns from databases. ACM SIGMOD Rec. **27**(2), 85–93 (1998)
6. Bécan, G., Acher, M., Baudry, B., Nasr, S.B.: Breathing ontological knowledge into feature model synthesis: an empirical study. Empir. Softw. Eng. **21**(4), 1794–1841 (2016)
7. Benali, K., Rahal, S.A.: Ontodta: ontology-guided decision tree assistance. J. Inf. Knowl. Manag. **16**(03), 1750031 (2017)
8. Bernecker, T., Kriegel, H.P., Renz, M., Verhein, F., Zuefle, A.: Probabilistic frequent itemset mining in uncertain databases. In: Proceedings of the 15th ACM SIGKDD International Conference on Knowledge Discovery and Data Mining, Paris, France, pp. 119–128 (2009)
9. Ceusters, W.: An information artifact ontology perspective on data collections and associated representational artifacts. In: Proceedings of the Medical Informatics in Europe Conference (MIE 2012), pp. 68–72 (2012)

10. Chang, J.H., Lee, W.S.: Finding recent frequent itemsets adaptively over online data streams. In: Proceedings of the 9th ACM SIGKDD International Conference on Knowledge Discovery and Data Mining, Washington, DC, USA, pp. 487–492 (2003)

11. Chi, Y., Wang, H., Yu, P.S., Muntz, R.R.: Moment: maintaining closed frequent itemsets over a stream sliding window. In: Proceedings of the 4th IEEE International Conference on Data Mining, Brighton, UK, pp. 59–66 (2004)

12. Chui, C.K., Kao, B., Hung, E.: Mining frequent itemsets from uncertain data. In: Pacific-Asia Conference on Knowledge Discovery and Data Mining, Nanjing, China, pp. 47–58 (2007)

13. Czarnecki, K., Hwan, C., Kim, P., Kalleberg, K.: Feature models are views on ontologies. In: 10th International Software Product Line Conference (SPLC 2006), Baltimore, MD, USA, pp. 41–51 (2006)

14. Giannella, C., Han, J., Pei, J., Yan, X., Yu, P.S.: Mining frequent patterns in data streams at multiple time granularities. Next Gener. Data Mining 212, 191–212 (2003)

15. Han, J., Pei, J., Yin, Y.: Mining frequent patterns without candidate generation. ACM SIGMOD Rec. 29(2), 1–12 (2000)

16. Hilario, M., Nguyen, P., Do, H., Woznica, A., Kalousis, A.: Ontology-based meta-mining of knowledge discovery workflows. In: Meta-Learning in Computational Intelligence, pp. 273–315 (2011)

17. Hong, T.P., Lee, C.H., Wang, S.L.: Mining high average-utility itemsets. In: IEEE International Conference on Systems Man and Cybernetics, San Antonio, TX, USA, pp. 2526–2530 (2009)

18. Hong, T.P., Lin, C.W., Wu, Y.L.: Incrementally fast updated frequent pattern trees. Expert Syst. Appl. 34(4), 2424–2435 (2008)

19. Jovanovska, L., Panov, P.: Semantic representation of machine learning and data mining algorithms. In: Proceedings of the 44th International Convention on Information, Communication and Electronic Technology (MIPRO), pp. 205–210 (2021)

20. Keet, C.M., et al.: The data mining optimization ontology. J. Web Semant. 32, 43–53 (2015)

21. Lee, V.E., Jin, R., Agrawal, G.: Frequent pattern mining in data streams. In: Aggarwal, C.C., Han, J. (eds.) Frequent Pattern Mining, pp. 199–224. Springer, Cham (2014). https://doi.org/10.1007/978-3-319-07821-2_9

22. Leung, C.K.S., Khan, Q.I.: Dstree: a tree structure for the mining of frequent sets from data streams. In: Proceedings of the 6th IEEE International Conference on Data Mining, Hong Kong, China, pp. 928–932 (2006)

23. Li, H.F., Lee, S.Y.: Mining frequent itemsets over data streams using efficient window sliding techniques. Expert Syst. Appl. 36(2), 1466–1477 (2009)

24. Li, H.F., Lee, S.Y., Shan, M.K.: An efficient algorithm for mining frequent itemsets over the entire history of data streams. In: Proceedings of 1st International Workshop on Knowledge Discovery in Data Streams (2004)

25. Manku, G.S., Motwani, R.: Approximate frequency counts over data streams. In: Proceedings of the 28th International Conference on Very Large Databases (VLDB), Hong Kong, China, pp. 346–357 (2002)

26. Panov, P., Soldatova, L., Džeroski, S.: Ontology of core data mining entities. Data Mining Knowl. Discov. 28(4), 1222–1265 (2014). https://doi.org/10.1007/s10618-014-0363-0

27. Panov, P., Soldatova, L.N., Džeroski, S.: Generic ontology of datatypes. Inf. Sci. 329, 900–920 (2016)

28. Pei, J., Han, J., Mao, R., et al.: Closet: an efficient algorithm for mining frequent closed itemsets. In: ACM SIGMOD Workshop on Research Issues in Data Mining and Knowledge Discovery, vol. 4, no. 2, pp. 21–30 (2000)

29. Publio, G.C., et al.: ML-schema: exposing the semantics of machine learning with schemas and ontologies. arXiv preprint arXiv:1807.05351 (2018)

30. Smith, B., et al.: The obo foundry: coordinated evolution of ontologies to support biomedical data integration. Nat. Biotechnol. **25**(11), 1251–1255 (2007)

31. Smith, B., et al.: Relations in biomedical ontologies. Genome Biol. **6**(5), 1–15 (2005)

32. Vanschoren, J., Blockeel, H., Pfahringer, B., Holmes, G.: Experiment databases. Mach. Learn. **87**(2), 127–158 (2012)

33. Yao, H., Hamilton, H.J., Butz, C.J.: A foundational approach to mining itemset utilities from databases. In: Proceedings of the 4th SIAM International Conference on Data Mining, Lake Buena Vista, Florida, USA, pp. 482–486 (2004)

34. Yu, J.X., Chong, Z., Lu, H., Zhou, A.: False positive or false negative: mining frequent itemsets from high speed transactional data streams. In: Proceedings of the 30th International Conference on Very Large Data Bases, Toronto, Canada, pp. 204–215 (2004)

35. Zaki, M.J.: Scalable algorithms for association mining. IEEE Trans. Knowl. Data Eng. **12**(3), 372–390 (2000)

Passing Heatmap Prediction Based on Transformer Model Using Tracking Data for Football Analytics

Yisheng Pei[1]([envelope]) [ID], Varuna De Silva[1] [ID], and Mike Caine[1,2]

[1] Institute for Digital Technologies, Loughborough University London, London, UK
y.pei2@lboro.ac.uk
[2] University of Warwick, Coventry, UK

Abstract. Although the data-driven analysis of football players' performance has been developed for years, most research only focuses on the on-ball event including shots and passes, while the off-ball movement remains a little-explored area in this domain. Players' contributions to the whole match are evaluated segmentally and unfairly, those who have more chances to score goals earn more credit than others, while the indirect and unnoticeable impact that comes from continuous movement has been ignored. This research presents a novel deep-learning network architecture which is capable to predict the potential end location of passes and how players' movement before the pass affects the final outcome. Once analyzed more than 28,000 pass events, a robust prediction can be achieved with more than 0.7 Top-1 accuracies. And based on the prediction, a better understanding of the pitch control and pass option could be reached to measure players' off-ball movement contribution to defensive performance. Moreover, this model could provide football analysts a better tool and metric to understand how players' movement over time contributes to the game strategy and final victory.

Keywords: Performance Analysis · Deep Learning · Football

1 Introduction

Data-based analysis of athletics performance has been developed for years, as the most complicated and popular sport around the world, the research on the performance of football players is being paid more and more attention [1–3]. Nowadays, with advances in technology and big data, we have several advanced metrics and methods to evaluate individual players' contribution to the match outcome or goal-scoring opportunities, for instance, xG [6], xT [27], and many other relatives [5,11]. While most previous research only focuses on the on-ball event during the competition, including passes [2,13], shot [14] and dribble, which ignores the fact that each player has only 3 min to control the ball on average as opposed to being free of the ball for 87 min. If players run randomly and spontaneously, the outcome will not likely be positive. Thus the decision-making

A. Bennour et al. (Eds.): ISPR 2023, CCIS 1940, pp. 162–173, 2024.
https://doi.org/10.1007/978-3-031-46335-8_13

capability during that 87 min without the ball determined whether they were really committed to the team's strategy and performance. This paper addresses the problem of estimating the movement decision capability when players are defending and propose a deep machine learning model that allows us to value and compare their decision with other potential choices.

The measurement of decision-making ability is essentially a comparison value between the happened decision of players during defending and the potential choice based on all the given spatiotemporal information. As the ultimate goal of attacking, the purpose of all the on-ball events relates to creating goal-scoring chances. On the other side of the coin, the purpose of defending is to avoid the increase of opponent teams' success probability. This paper takes the most frequent scene in a football match, passing, which covers over 80 per cent of the match events, as the research objective. To reach this level, we calculate the pass-end location possibilities and reflect the relationship with certain players' movement together.

We propose a self-attention mechanism model to learn the pass-end location possibilities depending on the defenders' trajectories. In order to achieve the purpose of valuing off-ball movement, we regard the problem as a time series classification task while the end zone of each pass is the category that is being predicted.

The main contributions of this work are the following:

- We propose a model for estimating the possible final outcome and the reality to evaluate football players' movement decision-making capability while defending, which allows us to analyse players on a different and comprehensive level and include more match context.
- We develop a self-attention architecture to estimate how players' movement affects the match process and shows that the deep learning model has much more potential for football performance analysis.
- We present a novel and practical metric which is able to help to identify outstanding defending players and improve their decision-making capability.

This paper is organised as follows: relevant literature in sports analysis and machine learning (ML) is presented in Sect. 2. The proposed xPass model is introduced in Sect. 3, along with all technical details. The results of experiments and potential application are shown in Sect. 4 and Sect. 5 respectively. Finally, in the last section, we summarise the paper and indicate the outlines of future work.

2   Background

Considering the practical and applied usage of xD metric, our work is inspired by several previous other approaches aimed at estimating [23,30,33] and predicting the game process based on the tracking [22,28] and event data in the professional football domain. Technically, instead of the traditional convolutional neural networks (CNN) in computer vision tasks, we leverage the latest self-attention transformer model [32] and treat the problem as a multi-class classification task.

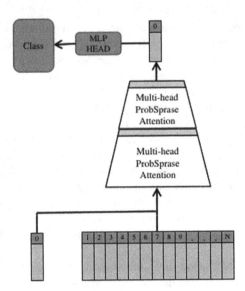

Fig. 1. The xPass Model Overview. The encoder receives sequential embedded tracking data (green series) with an extra learnable 'classification token' [8] for the classification task. Instead of the canonical self-attention block, we replace it with the ProbSparse self-attention from the Informer model [34] to avoid memory limitation. (Color figure online)

2.1   Football Performance Analysis

The performance analysis of football players has been developed since the Moneyball theory [16]. Afterwards, Rudd [23] firstly and creatively treated the football game as a Markov Chain, which quantifies and connects the actions of players with the goal-scoring probability. Since then various approaches have been built to achieve a similar goal, measure the contribution of players' on-ball events to the final game result [20]. Some of them treat the target straightly as the shot chances and quality, while others choose a more indirect way to measure the pitch control or heatmap on the pitch during the possession [21,24]. The technical methods applied in previous research differ, from the expert-guided development of algorithmic rules to a linear regression model with a set of handcrafted features [12], from a CNN-based model [4] to an assembled tree algorithm [10]. Little previous researches put attention to the other side of the coin, the majority of game playing and players' decision-making process, movement efficiency without the ball and how that choices discriminate smart players from normal ones.

2.2   Self-attention and Transformer

The transformer model is initially designed to solve the sequential machine translation task between two languages [32]. After the self-attention mechanism is approved to have strong feature representing and capturing capabilities, instead of recurrent neural network (RNN) series, it becomes the state-of-the-art in Natural Language Processing (NLP) domain. Based on its special characters, the transformer also shows a huge prospect in the image tasks including image classifications [8], object detection [19] and semantic segmentation [18].

Besides the single task, as the development of this architect, the multi-task problems can be solved as well, which shares similar features with football performance analysis. They both have various paired and unpaired inputs with a number of types and the outputs have a certain flexibility due to the specific requests. Some early research has been attempted to extract features from the event dataset and football broadcasting video [26,34] and recognize the group activities in team sports [17,29]. In our case, the transformer-based model is used to extract useful features from football tracking datasets and applied them to several practical tasks separately.

3   Methodology

Within this section, we propose a transformer-based time series deep learning model to estimate the possibility of the ball end location of any given pass based on prior information. To achieve the goal, the model is built on top of the tracking data collected from real football matches, containing the x and y location of all 22 players and the ball at 25 frames per second.

In architectural design, we used a combined classification and time stamp embedding to deal with the pass data while a ProbSparse attention encoder extracts features for further tasks. Our proposed model aims to solve the pass prediction problem as a time-series classification problem. Please refer to Fig. 1 for an overview and check the following sections for detail.

3.1   Loss Function

Defining a proper loss function that measures the gap between prediction and ground truth is necessary and crucial for a model. While the model output can be varied according to the tasks, not all the information is highly and directly associated with the target. For example, predicting the players who are top-3 most likely to receive the ball has been analysed before [2], and the number of players has been utilized as the target feature normally.

Instead of predicting the potential ball receiver, which is a 9-classes classification problem, the ball passer and the goalkeeper are ignored. Due to the reality of football matches, most of the passes are aimed at the spare space rather than a staying player. Thus the understanding of pass location possibility on the whole pitch matters, because it is much closer to the game philosophy. While the standard pitch is 106 m * 65 m, how to split the pitch into a large number of zones affects the final result significantly. For example, if we have either a single horizontal or vertical zone on the pitch like Fig. 2(A), the total amount would be up to 1200 which is too many for football datasets and the number

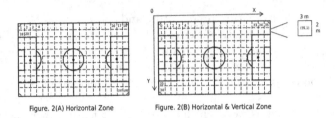

Figure. 2(A) Horizontal Zone   Figure. 2(B) Horizontal & Vertical Zone

Fig. 2. The whole football pitch is divided into multi zones on both the x-axis and y-axis. Each zone contains the same area of the original pitch.

will lose geographical meaning as zone 1 is actually next to zone 19, but would be marked as the neighbour of zone 18, which is on the other end of the pitch. The problem could be solved by using a combined coordinate on the x-axis and y-axis separately. Eventually, the pitch has two sides, one has 35 zones on the x-axis and 34 zones on the y-axis, leading to each zone being 3 m long and 2 m wide and can be labelled as $(35, 1)$ in Fig. 2(B), another one has 105 zones on the x-axis and 68 zones on the y-axis, leading to each zone being 1 m long and 1 m wide.

We train the xPass model by minimizing the Cross-Entropy Loss (CEL) between the estimated and ground truth on both the x-axis and y-axis. The loss function $L(\widehat{z}, z)$ is defined in Eq. 1.

$$L(\widehat{z}, z) = CEL(\widehat{z_x}, z_x) + CEL(\widehat{z_y}, z_y) \tag{1}$$

where $\widehat{z_x}$ and $\widehat{z_y}$ represents the prediction of the ball end location on the x-axis and y-axis separately while z_x and z_y means the ground truth.

3.2 Model Architecture

The xPass model comprises several stages according to different tasks. Inputs are the match tracking data from $(t - 2)$ to t seconds, with the output prediction of the pass event at time t.

Class Token and Global Embedding. The input representation process includes classification token embedding and time-series global embedding. The classification token embedding is inspired by the Vision Transformer work [8] and is alignable. While the time-series global embedding is required to capture the long-range sequential features and patterns for the task.

Similar to BERT's [class] token [7], the first token of every sequence is always a special classification token which would be extracted after the attention blocks. A classification head is attached afterwards. In Fig. 1, the 0 green series which was added at the beginning and extracted after the attention blocks stands for the classification token.

For the time-series global embedding, we have followed the method proposed by the Informer model [34]. Three separate parts are considered, a scalar projection, a local time stamp and a global time stamp. The scalar projection is aimed at aligning the dimension of inputs. While the local time stamp embedding is the same as the vanilla fixed position embedding, in order to solve the time sequential prediction problem, the global information (week, month and year) or other special time stamps like holidays and events would be considered. In the case of football, previous match results, recent fixtures, current goal differences etc. could be the factors.

After the input representation, the t-th sequence input X_t would be shaped into a matrix $X_{en}^t \in R^{(l_x+1)*d_{model}}$.

Efficient Self-attention Mechanism. Self-attention The canonical self-attention mechanism is initially proposed by the transformer paper [32]. It adopts the Query-Key-Value (QKV) matrix calculation, given the packed matrix representations of queries $Q \in R^{N*D_k}$, keys $K \in R^{M*D_k}$ and values $V \in R^{M*D_v}$. The classical scaled dot-product attention score function is as follows:

$$Attention(Q, K, V) = softmax(\frac{QK^T}{\sqrt{d_k}})V \tag{2}$$

where M and N represent the lengths of keys and queries; D_k and D_v denote the dimensions of keys and values; $A = Softmax(\frac{QK^T}{\sqrt{d_k}})$ is often called attention matrix; softmax is applied in a row-wise manner. The dot-products of queries and keys are divided by $\sqrt{D_k}$ to alleviate the gradient vanishing problem of the softmax function, which should be altered case by case.

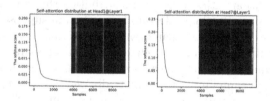

Fig. 3. The Softmax scores in the self-attention from a 4-layer canonical Transformer trained on the weather datasets [34].

One of the significant advantages of the self-attention mechanism compared to other time-series algorithms like Recurrent-Neural-Network (RNN) variants [25] is that the calculation can be fully parallelized, whereas an RNN always processes the sentences sequentially. The inherent cost of this mechanism leads to a higher demand for memory usage on GPU.

However, it has been proved that the distribution of self-attention probability has potential sparsity which means it forms a long tail distribution [31], and only a few dot-product pairs provide a contribution to the major attention score, while most of them are quick irrelevant (Fig. 3).

In a more visual and intuitionistic case, as shown in Fig. 4, at this specific pass event moment, due to the domain knowledge of professional football, the position of players in the red boxes has a much more direct and important impact on the ball holder's pass decision, while the players in the yellow boxes are supposed to affect less.

Fig. 4. The moment before a pass event happens on the pitch.

In order to distinguish the real contributed dot-product pairs from other lazy pairs, we follow the ProbSparse Self-attention mechanism from the Informer model [34]. The measurement consists of two parts. The first part is the query sparsity measurement, which compares the corresponding query's attention probability distribution with the uniform distribution through Kullbakc-Leibler (KL) divergence as the dominant dot-product pairs are supposed to differ from the uniform one while other lazy pairs tend to be similar. Second, based on the above query sparsity measurement, each key in the ProbSparse Self-attention is not attend to all queries but only to a specific amount of dominant pairs, which can be adjusted by a sampling factor. Eventually, the time complexity and space complexity of the ProbSparse Self-attention drops from $O(L^2)$ to $O(LlnL)$ significantly, where L stands for the input length of queries and keys.

The ProbSparse Self-attention equation is shown as:

$$Attention(Q, K, V) = softmax(\frac{\overline{Q}K^T}{\sqrt{d_k}})V \tag{3}$$

where $\overline{Q}$ is a sparse matrix of the same size as the original Q matrix, which only contains the top dominant queries rather than all of the inputs.

Encoder Stack for Sequential Inputs. The encoder is designed to extract robust and representative feature maps and is composed of a stack of $N = 2$ identical encoders. We give a sketch of the encoder in Fig. 5 for clarity.

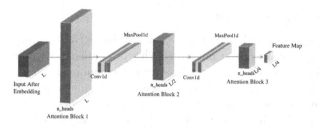

Fig. 5. The moment before a pass on the pitch.

Each encoder has $N = 3$ ProbSparse Attention Block and a distilling operation connects each of these attention blocks. Due to the initially designed purpose, only a few dominated queries contribute the most are selected, while the redundancy should be trimmed. The distilling operation consists of a 1-D Convolutional filter layer with a kernel width $= 3$, a Max-Pooling layer with stride 2 and an Exponential Linear Unit (ELU) activation function. The distilling operation forward from i-th distilling layer to $(i + 1)$-th layer is followed:

$$X_{i+1}^t = MaxPool(ELU(Conv1d([X_i^t]_{AB})))$$
(4)

where $[X_i^t]_{AB}$ is the t-sequence input X^t after the attention block. Since then the output dimension of the encoder stack is aligned, which is hinged on the number of encoder stacks and attention blocks. All the stacks' outputs would be concatenated at the end as the final output feature map of the encoder.

4 Experiment

In this section, we describe the datasets we used to train and validate our model with the testing performance of the proposed architecture for the pass-end location possibility estimation.

4.1 Datasets

This research is based on the tracking and event data generated from 30 Premier League matches of the 2019/2020 and 2020/2021 seasons, provided by a professional football club. The tracking data for each game contains the (x, y, z) location for each individual player and the ball at 25 Hz. The event data provides the time, event type, football team id, player involved, event start location, event end location, the outcome of the event and the body part used. In total, 16 different events are collected including pass, dribble, throw in and shot. From the datasets, we extract 28,872 passing events, which are split into a training, validation and test set with a 70:10:20 distribution. The validation set is used for model selection during a grid-search process while the result is based on the test dataset. Similar tracking and event dataset are available at https://github.com/metrica-sports/sample-data.

4.2   Experimental Details

We used the Adam optimizer with $1 = 0.9$ and $2 = 0.999$ for the proposed methods and started the learning rate from 1e–4 with a proper early stopping setting. The dimension of the transformer model is 512 and the numbers of the multi-head are 8. Two encoder stacks consist of the main encoder, while each of them has three ProbSparse self-attention blocks with two distilling connections. We trained the model on a NVIDIA RTX A6000 GPU with a batch size of 32.

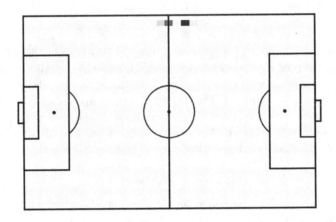

Fig. 6. The moment before a pass on the pitch.

We also compare our model with three previous related works [2,9,15]. The SoccerMap model is based on the heatmap of the tracking data with a combined computer vision and handcrafted features method to estimate the passing probability on the whole pitch, which consists of thousands of cells that approximately stand for $1\,m^2$ on the football pitch. We also cite two additional model results from their work, both of which are built on a set of handcrafted features generated from the tracking data. One is called *Logistic Net* which is a network with a single sigmoid unit, and another is *Dense2 Net* which is a neural network with two dense layers with a ReLu activation layer and a sigmoid output unit.

The *Pass Feasibility Model* is designed to predict the potential pass receiver based on the players' orientation, location and opponents' spatial configuration. They treat the problem as a 9-classes classification problem, excluding the ball holder and the goalkeeper. While another work *Pass Receiver Model* by [15], applied a computer vision method to the same task, comparing the difference between whether the visual information from the broadcasting video matters or not.

4.3   Results and Analysis

Table 1 presents the results of the proposed xPass model for the pass-end location prediction problem and three previous similar target works mentioned above.

Table 1. The result of the pass end location prediction with three relative models.

Model	CEL Loss	Top-1 Accuracy(%)	Top-3 Accuracy(%)	Top-5 Accuracy(%)
the xPass Model (2 * 3)	**0.249**	N/A	N/A	N/A
the xPass Model (1 * 1)	**0.277**	N/A	N/A	N/A
Logistic Net [9]	0.384	N/A	N/A	N/A
Dense2 Net [9]	0.349	N/A	N/A	N/A
SoccerMap [9]	**0.217**	N/A	N/A	N/A
Pass Feasibility Model [2]	0.624	37.6	71.0	N/A
Pass Receiver Model (without vision) [15]	0.510	49.0	84.9	95.0
Pass Receiver Model (with vision) [15]	0.375	62.5	92.3	97.5

Figure 6 shows the prediction heatmap of the passing end location, the darker colour zone represents a higher possibility to become the destination of the pass choice while the red point stands for the ground truth end location of that decision.

We convert the Top-1 Accuracy from [15] to the CEL loss as the comparison metric. However, their task is only a 9-classes classification problem while ours have up to 7,140 zones, which is on another level. Even though, our model overperforms those works, whether they are built on handcrafted features or a deep neural network with tracking and vision datasets. On the other hand, although the purpose of the SoccerMap model differs from ours, we both have a similar definition of the pitch zone and extend the pass event from a player-to-player relation but to a player-to-space event. We applied two totally different methods but match an even result, their work is based on the images generated from the tracking datasets and they focus on a certain passing moment while the time-series information before the moment is ignored. We believe that should not be missed whether from either a machine learning perspective or a football performance analysis perspective. That leads to the final result for us, without any handcrafted features but still achieving good performance on the task.

5   Conclusion

The novel xPass model applies the latest ProbSparse self-attention sequential machine learning techniques to predict the pass outcome and how it is affected by the other player's decision. The passing possibility heatmap is built on the model, which reflects players' off-ball movement contribution to defending during opponents' possession. In terms of practical application, the model has the capability of measuring the off-ball movement impact on the opponents' passing decisions and pitch control ability. The model based on the spatiotemporal tracking data can be transferred and applied to other similar sports, for many other tasks, it could also be the benchmark processing on the tracking and event data. We suggest more expert knowledge should be involved for a more detailed scenario analysis while more 3D body information will bring the model much closer and more realistic to the game situation.

References

1. Aalbers, B., Van Haaren, J.: Distinguishing between roles of football players in play-by-play match event data. In: Brefeld, U., Davis, J., Van Haaren, J., Zimmermann, A. (eds.) MLSA 2018. LNCS (LNAI), vol. 11330, pp. 31–41. Springer, Cham (2019). https://doi.org/10.1007/978-3-030-17274-9_3
2. Arbues-Sanguesa, A., Martin, A., Fernandez, J., Ballester, C., Haro, G.: Using player's body-orientation to model pass feasibility in soccer. In: 2020 IEEE/CVF Conference on Computer Vision and Pattern Recognition Workshops (CVPRW), pp. 3875–3884. IEEE Computer Society (2020). https://doi.org/10.1109/CVPRW50498.2020.00451
3. Bransen, L., Robberechts, P., Van, J., Davis, H.J.: Choke or shine? quantifying soccer players' abilities to perform under mental pressure (2019)
4. Cheong, L., Zeng, X., Tyagi, A.: Prediction of defensive player trajectories in NFL games with defender CNN-LSTM model (2021)
5. Decroos, T., Bransen, L., Haaren, J., Davis, J.: Vaep: an objective approach to valuing on-the-ball actions in soccer (extended abstract) (2020)
6. Decroos, T., Bransen, L., Van Haaren, J., Davis, J.: Actions speak louder than goals. In: Proceedings of the 25th ACM SIGKDD International Conference on Knowledge Discovery I& Data Mining (2019). https://doi.org/10.1145/3292500.3330758
7. Devlin, J., Chang, M.W., Lee, K., Toutanova, K.: Bert: pre-training of deep bidirectional transformers for language understanding (2018). https://arxiv.org/abs/1810.04805
8. Dosovitskiy, A., et al.: An image is worth 16×16 words: transformers for image recognition at scale (2022). https://openreview.net/forum?id=YicbFdNTTy
9. Fernández, J., Bornn, L.: SoccerMap: a deep learning architecture for visually-interpretable analysis in soccer. In: Dong, Y., Ifrim, G., Mladenić, D., Saunders, C., Van Hoecke, S. (eds.) ECML PKDD 2020. LNCS (LNAI), vol. 12461, pp. 491–506. Springer, Cham (2021). https://doi.org/10.1007/978-3-030-67670-4_30
10. Fernández, J., Barcelona, F., Bornn, L., Los, D., Dodgers, A.: Decomposing the immeasurable sport: a deep learning expected possession value framework for soccer (2019)
11. Fernández, J., Barcelona, F., Fernandez, J., Bornn, L.: Wide open spaces: a statistical technique for measuring space creation in professional soccer (2018)
12. Goes, F.R., Kempe, M., van Norel, J., Lemmink, K.A.P.M.: Modelling team performance in soccer using tactical features derived from position tracking data. IMA J. Manag. Math. **32**, 519–533 (2021). https://doi.org/10.1093/imaman/dpab006
13. Goes, F., Schwarz, E., Elferink-Gemser, M., Lemmink, K., Brink, M.: A risk-reward assessment of passing decisions: comparison between positional roles using tracking data from professional men's soccer. Sci. Med. Football **6**, 372–380 (2021). https://doi.org/10.1080/24733938.2021.1944660
14. Goes, F.R., Brink, M.S., Elferink-Gemser, M.T., Kempe, M., Lemmink, K.A.: The tactics of successful attacks in professional association football: large-scale spatiotemporal analysis of dynamic subgroups using position tracking data. J. Sports Sci. **39**, 523–532 (2020). https://doi.org/10.1080/02640414.2020.1834689

15. Honda, Y., Kawakami, R., Yoshihashi, R., Kato, K., Naemura, T.: Pass receiver prediction in soccer using video and players' trajectories. In: 2022 IEEE/CVF Conference on Computer Vision and Pattern Recognition Workshops (CVPRW), pp. 3502–3511 (2022). https://doi.org/10.1109/CVPRW56347.2022.00394

16. Lewis, M.: Moneyball: The Art of Winning an Unfair Game. Norton W.W, New York (2004)

17. Li, W., Yang, T., Wu, X., Du, X.J., Qiao, J.J.: Learning action-guided spatio-temporal transformer for group activity recognition. In: The 30th ACM International Conference on Multimedia (2022)

18. Lin, T., Wang, Y., Liu, X., Qiu, X.: A survey of transformers (2021)

19. Liu, Z., et al.: Swin transformer: Hierarchical vision transformer using shifted windows. In: 2021 IEEE/CVF International Conference on Computer Vision (ICCV), pp. 9992–10002. IEEE Computer Society, Los Alamitos (2021). https://doi.org/10.1109/ICCV48922.2021.00986. https://doi.ieeecomputersociety.org/10.1109/ICCV48922.2021.00986

20. Mackay, N.: Introducing a possession value framework (2018)

21. Martens, F., Dick, U., Brefeld, U.: Space and control in soccer. Front. Sports Active Living **3**, 676179 (2021). https://doi.org/10.3389/fspor.2021.676179

22. Robberechts, P., Van Haaren, J., Davis, J.: A bayesian approach to in-game win probability in soccer. In: Proceedings of the 27th ACM SIGKDD Conference on Knowledge Discovery I& Data Mining (2021). https://doi.org/10.1145/3447548.3467194

23. Rudd, S.: A framework for tactical analysis and individual offensive production assessment in soccer using Markov chains (2011)

24. Shaw, L., Gopaladesikan, S.: Routine inspection: a playbook for corner kicks. In: Brefeld, U., Davis, J., Van Haaren, J., Zimmermann, A. (eds.) MLSA 2020. CCIS, vol. 1324, pp. 3–16. Springer, Cham (2020). https://doi.org/10.1007/978-3-030-64912-8_1

25. Sherstinsky, A.: Fundamentals of recurrent neural network (RNN) and long short-term memory (LSTM) network (2018). https://arxiv.org/abs/1808.03314

26. Simpson, I., Beal, R.J., Locke, D., Norman, T.J.: Seq2event: learning the language of soccer using transformer-based match event prediction. In: Proceedings of the 28th ACM SIGKDD Conference on Knowledge Discovery I& Data Mining (2022). https://doi.org/10.1145/3534678.3539138

27. Singh, K.: Introducing expected threat (XT) (2018). https://karun.in/blog/expected-threat

28. StatsBomb: Introducing on-ball value (obv) (2021). https://statsbomb.com/2021/09/introducing-on-ball-value-obv/

29. Tamura, M., Vishwakarma, R., Vennelakanti, R.: Hunting group clues with transformers for social group activity recognition. arXiv:2207.05254 [cs] (2022)

30. Teranishi, M., Tsutsui, K., Takeda, K., Fujii, K.: Evaluation of creating scoring opportunities for teammates in soccer via trajectory prediction. arXiv:2206.01899 [cs] (2022)

31. Tsai, Y.H.H., Bai, S., Yamada, M., Morency, L.P., Salakhutdinov, R.: Transformer dissection: a unified understanding of transformer's attention via the lens of kernel (2019). arXiv:1908.11775. https://doi.org/10.48550/arXiv.1908.11775

32. Vaswani, A., et al.: Attention is all you need (2017)

33. Yam, D.: Attacking contributions: Markov models for football (2019)

34. Zhou, H., et al.: Informer: beyond efficient transformer for long sequence time-series forecasting. In: Proceedings of the AAAI Conference on Artificial Intelligence, vol. 35, pp. 11106–11115 (2021). https://doi.org/10.1609/aaai.v35i12.17325

Data Clustering Using Tangent Search Algorithm

Karim Bechiri[(✉)] and Abdesslam Layeb

Laboratory of Data Science and Artificial Intelligence (LISIA), NTIC Faculty, University Constantine 2 - Abdelhamid Mehri, Constantine, Algeria
{Karim.Bechiri,Abdesslem.layeb}@univ-constantine2.dz

Abstract. This article proposes a new algorithm, called Clustering Tangent Search Algorithm (C-TSA) to solve clustering problems. The C-TSA uses Tangent Search Algorithm (TSA) as an optimizer to move a given solution toward a better solution. The main characteristics of this new optimization algorithm are its simplicity, as it requires only a small number of user-defined parameters, and its efficiency in data clustering as demonstrated in the experimental results, where it provides very promising and competitive results on many benchmark datasets. TSA optimizer has been designed with an effective balance between exploration and exploitation. Additionally, it includes a useful local escape mechanism to prevent getting stuck in local optima, and an adaptive step size to improve convergence towards the best solution.

Keywords: Data mining · Clustering · Partitioning · Optimization · Unsupervised learning · Metaheuristic · Tangent Search Algorithm

1 Introduction

Data clustering is one of the most important techniques used in data mining, which involves recognizing and grouping similar multidimensional data vectors into a number of clusters or divisions. There are many uses of clustering algorithms in real world, such as data analysis and pattern recognition [1], knowledge discovery from the database in data mining [2] and either identifying groupings of houses based on their value, kind, and geographical location. Furthermore, clustering is widely used in computer vision in the pre-processing phase of image segmentation for medical imaging or biometric identification [3], and customer segmentation in market research [4]. Other uses of clustering include the classification of animals by identifying and grouping species following their similar genetic features [5], and even in earthquake research, where clustering can assist to analyze the next likely place where an earthquake might occur based on the areas impacted by an earthquake in a neighboring or similar region [6]. In [7], computational difficulties are discussed in cluster analysis and mathematical programming for optimal partitioning problems and algorithms efficiency.

Learning techniques are classified into two types: supervised and unsupervised methods. Supervised learning is a method of machine learning where an external teacher or

A. Bennour et al. (Eds.): ISPR 2023, CCIS 1940, pp. 174–188, 2024.
https://doi.org/10.1007/978-3-031-46335-8_14

supervisor provides the desired output for each input data set in order to train the model. This approach aims to learn a mapping between inputs and outputs, when given new inputs, the model can accurately predict their associated outputs. On the other hand, unsupervised learning does not have an external teacher or supervisor providing target classes for each data vector; it relies on sorting data vectors based on their distance.

In data mining literature, many clustering algorithms have been developed. The majority of them organize data into clusters regardless of the topology of the input space. In [8], clustering algorithms have been divided into 9 categories based on partition, hierarchy, density, and many others. There are several algorithms that belong to the partitioning approach; the well-known are K-Means (centroid-based), PAM (K-Medoids), and the CLARA algorithm for large datasets [8].

Centroid-based clustering is one type of unsupervised clustering where each cluster is represented by its center, known as the centroid. The centroid corresponds to the mean value of all points assigned to that particular cluster and it is updated on every algorithm iteration based on the values of elements in that class. This makes centroids an important tool for understanding how different clusters are related or distinct from each other in terms of their characteristics such as size, shape, density, etc. K-means algorithms, a centroid-based clustering, is one of the most popular and efficient clustering algorithms commonly used for partitioning a given data set into a set of (k) groups or clusters, where k represents the predetermined number of groups.

A variety of clustering algorithms uses heuristics or metaheuristics as an optimizer to move a given solution towards a better solution. Many of them are bioinspired, the first well-known is the Genetics Algorithm (GA), inspired by the process of natural selection, where evolution is made through crossover and mutation operators [9]. In the same family, another similar algorithm is Differential Evolution (DE) where the main idea is to generate a temporary individual based on individual differences within populations and then randomly restructure population evolutionary [10]. Particle Swarm Optimization (PSO) has become quite popular due to its proven performance. It simulates the social behavior of birds within a group, with each potential solution called particles forming into swarms [11].

In this paper, we propose a new centroid-based algorithm called the Clustering Tangent Search Algorithm (C-TSA), which uses a mathematical inspired metaheuristic called Tangent Search Algorithm (TSA) as an optimizer to move a solution in the direction of a better solution. We acknowledge that there are many different optimization algorithms that could be used for clustering, each with their own strengths and weaknesses. However, we chose TSA for several reasons. TSA has been shown to perform well on a wide range of optimization problems and provides very promising and competitive results on many benchmark functions [12]. Additionally, TSA is a relatively simple algorithm that requires only a small number of user-defined parameters, which makes it easy to implement for clustering problems. Lastly, TSA has not been studied in the context of clustering, so we believe that there is an opportunity to explore its potential in this area. To be fairer, in our study, we will compare the performance of C-TSA with the original, non-optimized versions of the most known clustering heuristics.

The paper includes the following sections: an overview of the Tangent Search Algorithm (TSA) in Sect. 2, explanation of adaptation details of TSA to clustering problem in Sect. 3, presentation and discussion of experimental results in Sect. 4, and the final conclusion will be drawn in Sect. 5.

2 Tangent Search Algorithm (TSA)

TSA is a population-based optimization algorithm, where each potential solution is called an agent. Initially developed to solve optimization problems, the TSA employs a mathematical model based on the tangent function to move a solution to a better one. The global moving step equations is named the tangent flight function. This step effectively balances exploration and exploitation search strategies, covering the search space well, as demonstrated by the author in [12]. Additionally, a novel escape procedure is used to avoid local minima. Moreover, this algorithm is equipped with an adaptive variable step size to enhance the convergence. The performance of TSA is confirmed by several tests: classical tests, CEC benchmarks, and engineering optimization problems.

The few user-defined parameters of TSA are *Pswitch, Pesc* and population *Size*. The first switching parameter, *Pswitch* ∈ [0,1], controls the balance between local and global random walks. A higher value of *Pswitch* indicates more intensification, which focuses on exploiting the local search space to refine the solutions. On the other hand, a lower value of *Pswitch* implies more exploration to find better solutions. Therefore, the choice of Pswitch value is crucial as it affects the trade-off between exploitation and exploration in the optimization process. The second parameter, *Pesc* ∈ [0,1], is the probability of applying the escape procedure. In the TSA, the limit of the search space related to the problem is identified by the lower bound (*Lb*) and upper bound (*Ub*), on each move, a solution is repaired if its value overflows *Lb* and *Ub* [12].

The TSA optimizer engine is described by the pseudo-code in Fig. 1.

```
While t <= MAX_FE // the maximum of function evaluations
   For each agent search Xi (i=1:T, T:size of the population)
      If rand < Pswitch
         Apply Intensification search;
      Else
         Apply Exploration search;
      End
   End for each
   If rand < Pesc
      Select randomly an agent search (G);
      Apply Escape local minima (G);
   End
   t=t+1;
End While
```

Fig. 1. The pseudo-code of TSA [12]

3   Clustering TSA (C-TSA)

This section presents the adaptation of the TSA algorithm to solve the clustering problem. In this case, various parameters must be defined, including the solution representation, search space, population size, as well as the *pswitch* and *pesc* parameters.

3.1   Representation of a Solution

In any centroid-based clustering algorithm, the goal is to find the most representative center for each class. In this case, a solution is a set of k-centers where k is the number of groups to create (dataset classes). Each center is represented by a vector where the dimension is the number of attributes in the dataset. A solution will have as many vectors as the number of desired classes, represented by a matrix of size [k, p], where k is the number of classes and p is the number of attributes as shown in Fig. 2.

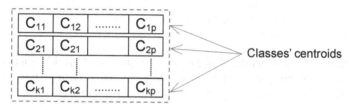

Fig. 2. Solution representation

3.2   C-TSA Parameters

First, like any optimization search algorithm, we have to define the search space delimited by a lower bound Lb and an upper bound Ub specific for each dimension (attribute). When loading the dataset to be partitioned, Lb and Ub are defined by extracting minimum and maximum values for each attribute. Additionally, based on an empirical study, TSA author recommends in [12] using a population size between 20 and 30. We chose arbitrary 30 as the value of population size in our experimental study. Finally, regarding the two parameters *Pswitch* and *Pesc,* we found by experimenting that the value of 0.5 for both parameters gives the best performances in the clustering context.

3.3   Evaluation Function

An evaluation function is a key component of any optimization algorithm. It allows us to measure the quality of a given solution and compare it with other solutions. In clustering problems, we try to find the most representative centers for each class in order to minimize the sum of all distances between elements and their centers according to the following equation:

$$\textbf{Fitness} = \text{Min}(\sum_{i=1}^{C}\sum_{j=1}^{Ni} Distance(Element_{ij}, Center_i)) \tag{1}$$

where C is the number of classes, Ni number of elements belonging to class i, $Element_{ij}$ represents the ith element of class j, and $Center_i$ is the center of class i.

The evaluation function is computed as follows: having a dataset of a size [t, p], where t is the number of elements and p is the number of attributes. For each element of the dataset, we make a Euclidean distance measurement with the k-centroids of a solution as shown in Fig. 3.

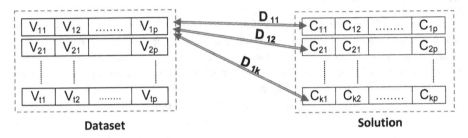

Fig. 3. Distance measurement

In Fig. 4, we obtain a matrix of distances with dimensions [t, k], where each row contains the distances between an element and the k-centroids. Then we make an assignment (classification) by associating each element to a class that is closest to the class center.

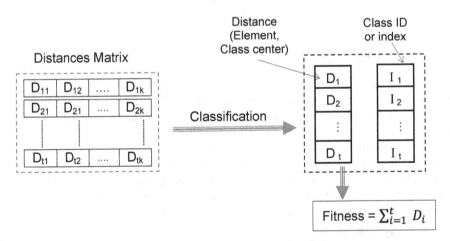

Fig. 4. Assignment and evaluation of a solution

Once all the elements have been classified, we can compute solution fitness by summing all the distances (elements with the centroid of their class). In our algorithm, the objective is to minimize as much as possible this value, in this case, the *Bestsolution* is the solution with the smallest fitness.

3.4 Flow Chart of the Proposed Approach C-TSA

Our algorithm begins by generating the initial population with random values within the limits of the search space [Lb$_i$, Ub$_i$] specific to each dimension. Then the iterations of the algorithm begin. In each iteration, we move (modify) all the solutions either by the intensification or the exploration operator according to *Pswitch* value. Then, we evaluate the obtained solution and update the best solution *Bestsolution*. The escape procedure is applied or not depending on a random value, if so, we evaluate and update *Bestsolution*. These steps are repeated until the stop criterion is met. Algorithm flow chart is presented in Fig. 5.

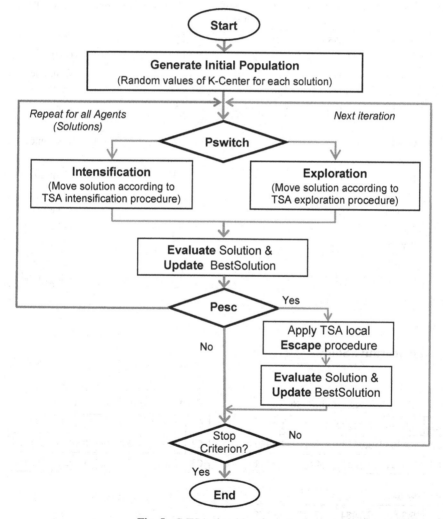

Fig. 5. C-TSA algorithm flow chart

4 Experimentation and Discussion

The proposed algorithm is implemented in the MATLAB R2017a environment, and all the experiments were performed on a computer equipped with Windows 10/64-bit system, Intel i5 processor (2.9 GHz), and 16 GB of RAM. For comparison and validation study, we used as benchmark 8 different public data collections, downloaded from [13] and [14]. The definition of each dataset is presented in Table 1.

Table 1. Datasets description

ID	Name	Number of cases	Number of attributes	Number of classes	Data type
1	Balance	625	4	3	Discrete, categorical
2	Cancer	699	9	2	Continue, real
3	Ecoli	336	7	8	Continue, real
4	Glass	214	9	2	Continue, real
5	Iris	150	4	3	Continue, real
6	Ovarian	216	100	2	Continue, real
7	Thyroid	7200	21	3	Mixed
8	Covid_19	9778	8	3	Discrete, categorical

4.1 Comparative Study

C-TSA   Fitness = 96,6686

C1	5,9185	2,7922	4,4100	1,4149
C2	6,7364	3,0678	5,6142	2,0960
C3	5,0121	3,4031	1,4717	0,2354

PSO   Fitness = 96,6731

C1	5,9260	2,7946	4,4039	1,4116
C2	6,7274	3,0623	5,6426	2,1078
C3	5,0166	3,3992	1,4758	0,2318

K-Means   Fitness = 97,3259

C1	5,9016	2,7484	4,3935	1,4339
C2	6,8500	3,0737	5,7421	2,0711
C3	5,0060	3,4180	1,4640	0,2440

Fig. 6. The best solutions found by C-TSA, PSO, and K-means for iris dataset

To show the efficiency of C-TSA, we made a comparison with others existing algorithms, such as PSO, DE, GA and the famous K-Means. The source codes of those algorithms are available in [15]. For a fair comparison during this experiment, the solutions are evaluated with the same Fitness function. We set the population size at 30 search agents for all algorithms. For example, after running 100 iterations of the above algorithms with the Iris dataset, we obtained in Fig. 6 results for the best solutions.

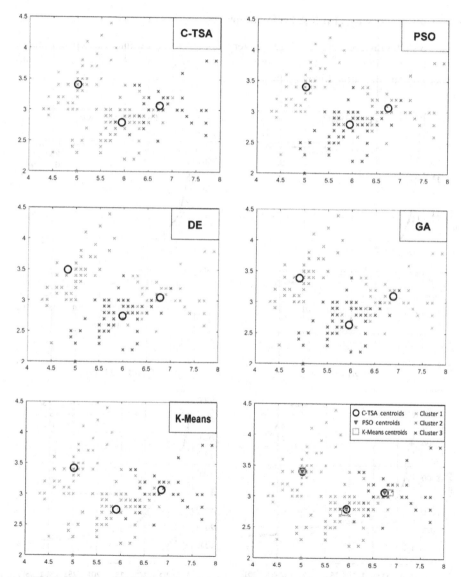

Fig. 7. Results of C-TSA, PSO, DE, GA and K-Means clustering on the Iris dataset

We can see in Fig. 6 the effectiveness of C-TSA which was able to find the best Fitness more accurately than the PSO algorithm and even better than the K-Means.

Figure 7 presents a graphic representation of the optimal solutions found by the different clustering algorithms, the colored points are the elements of the three classes, and the 3 circles represent the centroids of each class. The bottom right graphics compare the best solutions found by C-TSA, PSO and K-means.

4.2 Convergence and Fitness Performance

Figure 8 shows the efficiency of C-TSA in terms of convergence and fitness performance versus the different clustering algorithms for different datasets. An average of 10 runs is used to draw the comparison graphs.

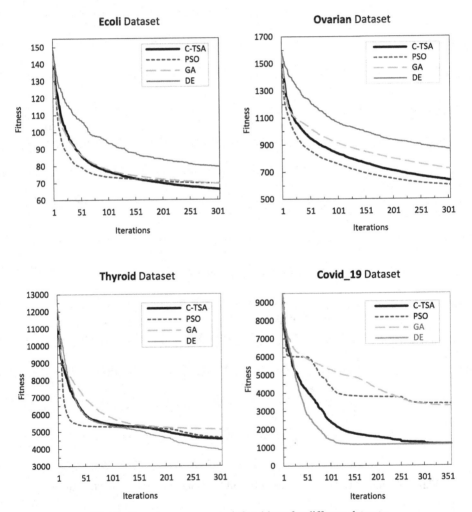

Fig. 8. Convergence curves of algorithms for different datasets

The experimental results presented in Fig. 8 clearly demonstrate the efficacy of C-TSA in dealing with the clustering of various datasets. C-TSA is adaptable to different types of data collections, and outperforms GA and DE for all datasets except Thyroid and Covid_19. In addition, C-TSA exhibits comparable convergence performance to PSO overall, except for Covid_19 where it demonstrates significantly better performance.

4.3 Evaluate Clustering Quality

In this section, we measure the effectiveness of C-TSA against these competing clustering algorithms by using partitioning quality measures, such as Accuracy and the Rand index.

Generally, the Accuracy is used to measure the quality of clustering or classification. It is computed as the sum of the diagonal elements of the confusion matrix, divided by the number of samples to obtain a value between 0 and 1. Its formula is described as:

$$Accuraccy(y, \hat{y}) = \frac{1}{n} \sum_{i=0}^{n-1} 1(\hat{y}_i = y_i) \tag{2}$$

where n represent number of samples in the dataset, as well as y_i and $\hat{y}_i$ represents the real classes and predicted classes of the ith element, respectively. When y_i and $\hat{y}_i$ are equal, expression took 1, otherwise 0.

The second measure used is the Rand Index (RI), computes the similarity of the results between two different clustering methods, in our case the measurement is made between the found clusters and the real groupings. RI is calculated as follows:

$$RI = \frac{(number\ of\ matching\ pairs)}{(number\ of\ pairs)} \tag{3}$$

The Rand index always takes a value between 0 and 1. Where, 0 means that the two clustering methods do not agree on the clustering of a pair of elements, and 1 means that they agree perfectly on the clustering of each pair.

Now, let's move to the presentation of the results of our experiments. Each test run is repeated 30 times and the maximum number of function evaluations is set at 10000. The experimental results are summarized in Tables 2, 3 and, 4. The best results are in bold.

Table 2. Statistic results of the Fitness (*the lowest is the best*)

		C-TSA	PSO	DE	GA	K-Means
Balance	best	1 423,82	1 423,82	1 426,67	1 423,82	1 423,85
	worst	1 432,84	1 431,85	1 431,35	1 433,40	1 433,03
	mean	1 425,80	**1 424,56**	1 428,67	1 425,99	1 426,24
	Std	1,72	1,57	1,09	1,85	2,74

(continued)

Table 2. (*continued*)

		C-TSA	PSO	DE	GA	K-Means
Cancer	best	303,53	303,53	303,53	303,53	305,93
	worst	303,53	303,53	304,08	303,59	306,08
	mean	**303,53**	**303,53**	303,68	303,55	306,07
	Std	0,00	0,00	0,16	0,01	0,04
Ecoli	best	63,69	64,91	74,62	64,06	62,41
	worst	68,79	72,22	82,17	74,49	66,19
	mean	65,79	68,90	78,95	69,62	**64,03**
	Std	1,15	1,91	1,99	2,26	1,11
Glass	best	311,42	311,42	311,42	311,43	318,65
	worst	311,43	395,83	311,90	395,86	428,72
	mean	**311,42**	314,24	311,50	314,29	333,36
	Std	0,00	15,41	0,13	15,41	31,05
Iris	best	96,66	96,66	96,67	96,66	97,33
	worst	96,67	127,67	100,90	97,50	124,18
	mean	**96,66**	97,69	98,30	97,03	100,79
	Std	0,01	5,66	1,22	0,36	8,95
Ovarian	best	580,81	566,67	640,46	635,68	526,11
	worst	662,07	743,99	955,95	808,97	526,78
	mean	621,85	605,94	826,34	728,72	**526,53**
	Std	20,15	44,62	87,69	48,30	0,33
Thyroid	best	3 496,84	3 866,37	3 296,11	4 587,82	3 470,22
	worst	4 796,95	5 284,18	5 264,45	5 287,13	5 175,66
	mean	4 147,83	4 783,95	4 008,35	5 218,89	**3 991,48**
	Std	348,42	384,53	425,18	134,06	557,18

4.4 Discussion of the Results

The comparison study of C-TSA shows generally a good stability in the majority of the datasets. We note that the C-TSA, generally has a low standard deviation of Fitness, Accuracy, and Rand Index compared to other algorithms, demonstrating its likely robustness. We noted that the Balance dataset is a difficult test for all algorithms without exception, the Accuracy and Rand Index values are quite low, as seen in Tables 3 and 4. In general, the C-TSA gives very good results of Rand Index and Accuracy except for the Thyroid dataset, demonstrating a lower performance compared to PSO and GA. Results in Table 3 show that K-means and DE are not efficient with the Thyroid dataset.

Overall, it is evident that there is no one algorithm that clearly stands out from the rest, but it is clear that the C-TSA ranks well among its competitors. The nature of

Table 3. Statistic results of the Accuracy (*the greatest is the best*)

		C-TSA	PSO	DE	GA	K-Means
Balance	best	62,87	65,54	62,60	62,65	64,88
	worst	54,20	56,88	54,66	53,81	54,75
	mean	58,89	**59,25**	58,61	58,08	58,71
	Std	1,85	1,73	1,96	1,93	1,95
Cancer	best	92,83	92,83	92,83	92,83	92,04
	worst	92,83	92,83	92,56	92,83	91,77
	mean	**92,83**	**92,83**	92,82	**92,83**	91,79
	Std	0,00	0,00	0,05	0,00	0,07
Ecoli	best	90,17	89,00	89,19	89,69	83,42
	worst	79,20	80,20	78,26	81,43	77,71
	mean	83,46	83,85	83,61	**85,94**	81,07
	Std	3,19	2,86	2,92	2,25	1,49
Glass	best	82,98	82,98	82,98	82,98	81,47
	worst	79,99	64,53	79,99	64,53	62,56
	mean	**82,65**	81,94	82,18	81,76	77,13
	Std	0,85	3,43	1,22	3,41	6,84
Iris	best	88,59	88,59	91,24	91,24	87,97
	worst	87,97	76,77	87,37	87,97	71,43
	mean	88,47	88,20	88,83	**89,76**	85,49
	Std	0,25	2,16	1,23	1,27	5,54
Ovarian	best	68,53	70,87	66,32	64,25	71,48
	worst	61,86	50,87	49,77	49,78	69,68
	mean	64,14	65,27	55,07	56,47	**70,34**
	Std	1,78	4,27	5,57	5,07	0,88
Thyroid	best	67,83	86,03	81,22	85,98	76,81
	worst	44,20	54,29	44,20	54,61	43,33
	mean	57,34	69,24	53,85	**80,68**	52,75
	Std	6,17	8,43	7,12	6,40	9,70

the data greatly influences the performance of each algorithm. These observations are confirmed by a statistical test, as you can see in Table 5, a repeated measures analysis of variance (ANOVA) with a Greenhouse-Geisser correction, which determined that the results of the different algorithms did not differ significantly for the three-evaluation metrics: Fitness, Accuracy, and Rand Index.

Table 4. Statistic results of the Rand Index (*the greatest is the best*)

		C-TSA	PSO	DE	GA	K-Means
Balance	best	58,24	59,04	57,44	58,40	58,40
	worst	44,00	48,96	46,24	42,24	45,60
	mean	52,13	**53,00**	51,69	50,95	51,57
	Std	2,84	1,91	3,11	3,16	2,68
Cancer	best	96,28	96,28	96,28	96,28	95,85
	worst	96,28	96,28	96,14	96,28	95,71
	mean	**96,28**	**96,28**	**96,28**	**96,28**	95,72
	Std	0,00	0,00	0,03	0,00	0,04
Ecoli	best	82,44	82,44	81,25	81,55	68,75
	worst	51,19	54,76	49,70	61,61	47,02
	mean	66,70	67,59	66,59	**72,85**	57,81
	Std	8,58	8,03	7,28	5,28	5,23
Glass	best	90,65	90,65	90,65	90,65	89,72
	worst	88,79	77,10	88,79	77,10	75,23
	mean	**90,45**	89,94	90,16	89,83	86,54
	Std	0,53	2,50	0,77	2,48	5,22
Iris	best	90,00	90,00	92,67	92,67	89,33
	worst	89,33	66,67	88,67	89,33	51,33
	mean	89,87	89,22	90,22	**91,18**	84,31
	Std	0,27	4,26	1,27	1,29	12,07
Ovarian	best	80,56	82,41	78,70	76,85	82,87
	worst	74,54	57,41	50,00	50,93	81,48
	mean	76,70	77,31	62,99	66,44	**81,99**
	Std	1,62	5,13	9,92	8,10	0,68
Thyroid	best	80,03	92,58	89,65	92,56	86,78
	worst	54,65	64,57	54,53	64,86	51,90
	mean	68,05	80,25	64,07	**88,92**	63,72
	Std	6,95	7,50	7,09	5,31	10,43

Table 5. ANOVA with a Greenhouse-Geisser correction, for the three-evaluation metrics

Accuracy	Best	$F(1.411, 8.456) = 0.880$	$p = 0.413$
	Worst	$F(1.411, 8.456) = 0.880$	$p = 0.413$
Rand Index	Best	$F(1.312, 7.869) = 0.993$	$p = 0.375$
	Worst	$F(1.632, 9.791) = 0.494$	$p = 0.588$
Fitness	Best	$F(1.312, 7.869) = 0.993$	$p = 0.375$
	Worst	$F(1.632, 9.791) = 0.494$	$p = 0.588$

5 Conclusion

In this paper, we have presented the C-TSA, a new clustering method based on the TSA algorithm. The performance of this approach has been compared with other popular methods. The tests were carried out on eight public databases. The experimental results showed the good performance of the new C-TSA method and its ability to make good quality data clustering compared to other existing algorithms. Furthermore, in most cases the proposed method outperformed the traditional K-Means. Moreover, the statistical study proved the effectiveness of the TSA adaptation to the clustering problem, because its performances are similar to those of the popular and robust clustering metaheuristics such as GA, PSO and DE.

While C-TSA has shown promising performance on various benchmark datasets, it also exhibits some weaknesses, as seen on the Thyroid dataset, its Rand Index and accuracy are lower than those of GA, which requires further investigation. Hybridizing C-TSA with local search heuristics could improve its performance, especially on more complex datasets. Additionally, exploring C-TSA's application in real-world clustering problems like market segmentation, fraud detection, or bioinformatics would be valuable to assess its practical performance. Improving C-TSA's robustness in diverse scenarios and datasets would enhance its effectiveness as a clustering algorithm.

References

1. Ezhilmaran, D., Vinoth Indira, D.: A survey on clustering techniques in pattern recognition. AIP Conf. Proc. **2261**, 030093 (2020). https://doi.org/10.1063/5.0017774
2. Varun, K., Nisha, R.: Knowledge discovery from database using an integration of clustering and classification. Int. J. Adv. Comput. Sci. Appl. **2** (2011). https://doi.org/10.14569/IJACSA.2011.020306
3. Mittal, H., Pandey, A.C., Saraswat, M., et al.: A comprehensive survey of image segmentation: clustering methods, performance parameters, and benchmark datasets. Multimed Tools Appl. **81**, 35001–35026 (2022). https://doi.org/10.1007/s11042-021-10594-9
4. Ramasubbareddy, S., Srinivas, T.A.S., Govinda, K., Manivannan, S.S.: Comparative study of clustering techniques in market segmentation. In: Saini, H., Sayal, R., Buyya, R., Aliseri, G. (eds.) Innovations in Computer Science and Engineering. LNNS, vol. 103. Springer, Singapore (2020). https://doi.org/10.1007/978-981-15-2043-3_15

5. Ramaraj, E., Punithavalli, M.: Taxonomically clustering organisms based on the profiles of gene sequences using PCA. J. Comput. Sci. **2**(3), 292–296 (2006). https://doi.org/10.3844/jcssp.2006.292.296

6. Yuen, D.A., Dzwinel, W., Ben-Zion, Y., Kadlec, B.: Earthquake clusters over multidimensional space, visualization of. In: Meyers, R. (eds.) Encyclopedia of Complexity and Systems Science. Springer, New York (2009). https://doi.org/10.1007/978-0-387-30440-3_145

7. Hansen, P., Jaumard, B.: Cluster analysis and mathematical programming. Math. Program. **79**, 191–215 (1997). https://doi.org/10.1007/BF02614317

8. Xu, D., Tian, Y.: A comprehensive survey of clustering algorithms. Ann. Data. Sci. **2**, 165–193 (2015). https://doi.org/10.1007/s40745-015-0040-1

9. Murty, M.N., Rashmin, B., Bhattacharyya, C.: Clustering based on genetic algorithms. In: Ghosh, A., Dehuri, S., Ghosh, S. (eds.) Multi-Objective Evolutionary Algorithms for Knowledge Discovery from Databases. Studies in Computational Intelligence, vol. 98. Springer, Heidelberg (2008). https://doi.org/10.1007/978-3-540-77467-9_7

10. Chen, G., Luo, W., Zhu, T.: Evolutionary clustering with differential evolution. In: IEEE Congress on Evolutionary Computation (CEC), Beijing, China, pp. 1382–1389 (2014). https://doi.org/10.1109/CEC.2014.6900488

11. Abraham, A., Das, S., Roy, S.: Swarm intelligence algorithms for data clustering. In: Maimon, O., Rokach, L. (eds.) Soft Computing for Knowledge Discovery and Data Mining. Springer, Boston (2008). https://doi.org/10.1007/978-0-387-69935-6_12

12. Layeb, A.: Tangent search algorithm for solving optimization problems. Neural Comput& Appl. **34**, 88538884 (2022). https://doi.org/10.1007/s00521-022-06908-z

13. Dua, D. and Graff, C.: UCI Machine Learning Repository. Irvine, CA: University of California, School of Information and Computer Science (2019). http://archive.ics.uci.edu/ml

14. Yazeed Zoabi (2020). https://github.com/nshomron/covidpred

15. Heris, M.K.: Evolutionary Data Clustering in MATLAB, Yarpiz (2015). https://yarpiz.com/64/ypml101-evolutionary-clustering

Multiple Diseases Forecast Through AI and IoMT Techniques: Systematic Literature Review

Asma Merabet[1,3](✉), Asma Saighi[1,3], Zakaria Laboudi[1],
and Mohamed Abderraouf Ferradji[2]

[1] University of Larbi Ben M'hidi Oum El Bouaghi, Oum El Bouaghi, Algeria
{asma.merabet,saighi.asma,laboudi.zakaria}@univ-oeb.dz
[2] University of Ferhat Abbes Setif, Setif, Algeria
mohamed.ferradji@univ-setif.dz
[3] LIAOA Laboratory, University of Oum El Bouaghi, Oum El Bouaghi, Algeria

Abstract. Over the past few years, researchers and developers have managed to overcome several challenges in order to provide informative, interactive and effective healthcare solutions. In particular, the recent developments in Artificial Intelligence (AI) field, more specifically ML and DL techniques, have contributed significantly to making Clinical Decision Support Systems (CDSS) more effective in healthcare processes by improving diagnostics, therapy, and prognosis. On another side, the Internet of Medical Things (IoMT), which has evolved into a tool to next-generation bioanalysis., combines networked biomedical devices with software applications to efficiently support healthcare tasks. Practically speaking, persons are susceptible to suffer from one or more chronic or non-chronic diseases under several conditions. This is why AI and IoMT are believed to enable the early identification of potential threats to human health that require effective health actions. In this paper, we accomplish an SLR of AI-based CDSS and IoMT techniques for multi-disease forecasting by making analysis and discussions according to various aspects. The aim is to help researchers in this field of interest to open up future prospects, especially since the existing literature reviews on medical decision support systems mainly focus on the prediction of a single disease rather than multiple diseases.

Keywords: Clinical Support System · Machine Learning · Deep Learning · IoMT · Systematic literature review

1 Introduction

Healthcare systems have been majorly revolutionized over the last years due to the tremendous progress in digital health technologies such as artificial intelligence (AI), robotics, IoMT, etc. This advancement has noticeably reduced the medical diagnostic and monitoring errors made by humans due to tiredness,

A. Bennour et al. (Eds.): ISPR 2023, CCIS 1940, pp. 189–206, 2024.
https://doi.org/10.1007/978-3-031-46335-8_15

work overload, and massive generated data [1], Consequently clinical outcomes can be improved, and data can be tracked over time through digital health. Artificial Intelligence (AI) and IoMT technologies are increasingly being used in the Healthcare sector, especially in advanced clinical decision support systems (CDSS) applications. It provides solutions that would assist in accelerating the decision-making processes within healthcare systems for precisely, earlier, and more reliably focused medical treatments [2]. IoMT has a significant impact on data collection with patient data monitoring, whereas AI is expected to analyze the increasing amounts of data and make decisions according to what it learns from the data. The most important factor in the treatment of any disease is to predict or recognize early diseases. According to [2], in the United States, it is predicted that there will be 1.9 million new cancer diagnoses and 609,360 cancer deaths in 2023. Through early identification of a patient's risk of cancer, along with other clinically relevant information, predictive models with AI using routinely performed blood tests have the potential to help physicians diagnose and deliver effective treatment to cancer patients earlier [3]. DL and ML are the areas that can be used to support the prediction of data-driven diagnosis systems. Researchers have introduced several ML and DL models to tackle the problem of huge and various data in promoting intelligent disease diagnosis to diagnose various diseases. Such models may predict the early diagnosis of the disease and provide solutions. Early diagnosis and efficient treatment are the best solutions to minimize the mortality rates caused by any disease. Consequently, the majority of medical experts become attracted to the new predictive models for disease prediction built around machine learning algorithms and deep learning. [4]. With the occurrence and emergence of telemedicine and intelligent health care came a major problem of restricted data access and limited data access. Therefore the IoMT is attracting the interest of the healthcare research community, where its potential to generate disruptive healthcare innovations may play a key role in reducing the pressure within health systems. Besides, combining IoMT with AI may deliver tailored health services that participate directly in enhancing the patient's life quality [5]. Our principal research aim is to address The following questions using SLR guidelines.

1. How effective is Artificial Intelligence in medical decision support systems ?
2. Which diseases were treated using Machine Learning and Deep Learning ?
3. How did the researchers implement the IoMT to predict disease?
4. What motivates using artificial intelligence in medical decision-supporting systems ?

This paper presents an SLR of approaches and techniques for AI-based medical decision support systems. The main contributions of this SLR include Multiple diseases predictions using machine learning, deep learning, and IoMT. The rest of the paper is structured as follows: Sect. 1 presents a Background about the used technologies, Sect. 2 presents Material and Methods. The Sect. 3 is for Results and Interpretation, and we conclude the paper with general discussion and perspectives.

2 Background

2.1 Machine Learning

According to Arthur Samuel [6], ML is the area of study that enables computers to learn autonomously, without requiring explicit programming. It is used for training machines to handle data more efficiently [6]. ML implements complex algorithms to recognize patterns in large amounts of data and predict outcomes independently of specific codes. We can classify ML into three classes: unsupervised, supervised, semi-supervised, and reinforcement learning. Supervised learning is where ML algorithms are trained on labeled data. Semi-supervised learning techniques have the ability to train machine learning models using both labeled and unlabeled data. Meanwhile, unsupervised learning algorithms aim to identify natural connections and patterns within unlabelled data. Reinforcement learning, on the other hand, is a general concept for ML approaches that include both prediction and decision-making. Such ML technology has an iterative approach to learning and can adapt based on initial feedback [7]. The Table 1 presents the ML algorithms used in this systematic review:

2.2 Deep Learning

A DL architecture is an artificial neural network (ANN) having two or more hidden layers to achieve higher prediction accuracy [10]. Deep learning applications in health care cover various problems, such as cancer detection, disease monitoring, and individual treatment advice. Deep learning works through learning models in data structures with neural networks of multiple convolution nodes of artificial neurons [11].

2.3 Internet of Medical Things (IoMT)

The term IoMT is used to describe the interconnection of medical devices. Devices that communicate effectively with one another and integrate into larger-scale healthcare systems to improve patient health [12]. IoMT devices allow for healthcare monitoring without the need for human intervention through the integration of automation, interfacing sensors. IoMT technology enables patients to remotely connect with clinicians, granting them access to medical care from a distance and transfer medical data over a secure network [13].

3 Material and Methods

3.1 Search Strategy

A systematic literature review defines the available technologies and approaches applied to evaluate the clinical efficacy of CDS systems in disease detection and prediction. The references used in the review study were found through searches of papers in Google Scholar, IEEE Xplore Digital Library, Springer,

Table 1. ML Algorithms Summary.

ML Algorithm	Type	Description	References:
K Nearest Neighbors	supervised	It is mainly used in classification problems. KNN ranks the objects according to the nearest distance, i.e., the proximity between the item and all the other items in the learning data	[20]
Support Vector Machine	supervised	The support vector machine is mainly used for regression and classification. It finds the optimal hyperplane for data classification.	[7]
Decision Tree	supervised	Decision trees are used to predict the dependent variable values by learning from simple decision rules derived from the data. It is formed with criteria defined as "entropy"	[19]
Random Forest	Unsupervised	The RF classifier is composed of multiple decision trees of the individual items in the given data set, it picks the average of the sub-set of each tree. Each node decision tree executes a query about the data.	[20]
Logistic Regression (LR)	Unsupervised	The LR models are acquired from the statistics branch for binary classification. The model LR chooses the class's likelihood of given data instances to predict as 0 or 1.	[20]
K-means	Unsupervised	In this approach, the dataset is partitioned into K clusters, where each cluster is represented by the mean value of its constituent samples. This mean value is commonly referred to as the "centroid."	[7]
Naïve Bayes	Unsupervised	Naïve Bayes is a robust machine-learning algorithm to predict outcomes. We use it to choose the best hypothesis (h) that fits the given data (d).	[8]
Artificial Neural Network	Supervised	The ANNs have neurons interconnected to each other. They are connected within the layers of the network. It can solve problems that were impossible by human or statistical standards. The ANN contains input hidden, and output layers.	[21]
C5.0	Supervised	C5.0 is an example of a decision tree algorithm used in ML. It build a model by using the tree structure based on relationship between features and potential outcomes. It can manage both nominal and numerical features. C5.0 is a ML algorithm that is based on C4.5, a classification algorithm which uses the concept of entropy of the information to find the feature impurity.	[9]

and Elsevier, including a combination of ML keywords (Artificial Intelligence, Disease Prediction, machine learning, IoMT, Deep learning, and medical decision support systems). The articles used in this research consist of papers from 2018.

3.2   Identify the Research Questions

The Table 2 the motivations of our research:

Table 2. Description of the motivations behind the different research questions.

Research Question	Motivation
1. How effective is Artificial Intelligence in medical decision support systems?	1. There are many research types to predict several diseases using AI algorithms. AI is important for healthcare as it can be associated with other technologies such as IoMT, cloud, and big data to achieve better results.
2. Which diseases were treated using ML and DL ?	2. Heart disease, diabetes, brain tumor, COVID19, Alzheimer
3. How did the researchers implement the IoMT to predict disease?	3. As data is the most crucial part of improving disease prediction models, some researchers used sensors to gather data to test the IoMT-based model.
4. What motivates using artificial intelligence in medical decision-supporting systems?	4. Artificial intelligence provides doctors with a quick and efficient pre-treatment and disease prediction decision-making process

3.3 Study Selection and Data Used in the Articles

Researchers demonstrated the effectiveness of the new technologies; they applied innovative systems built on machine learning and deep learning, in addition to IoMT. They used different algorithms to demonstrate the benefit of artificial intelligence along with their methodologies. To apply the different algorithms, they have used various data gathered from The Cleveland heart disease dataset 2016 which is publicly available from the University of California, Slides of human cancer tissue from the NCT biobank and the UMM pathology, (the Genomic Data Commons Data Portal (TCGA Pan-Cancer Clinical Data Resource), (Shandong Provincial Hospital dataset), (the Radiological Society of North America (RSNA)), Some researchers have encountered a few data sets, joined them, and built new one to work on (the Cleveland, Hungary, Switzerland, VA Long Beach, and Starlog heart disease dataset).

4 Results and Interpretation:

We divided this section into 2 parts, First, we presents researches applying Machine Learning algorithms to diagnose diseases with IoMT. The second part includes and discusses studies using deep learning and IoMT to predict diseases.

4.1 Machine Learning and IoMT-Based Diagnostic Applications

ML is increasingly used across various fields, including diagnosing diseases in healthcare. Many researchers have offered decision support systems for diagnosis based on machine learning [15]. In [14] researchers proposed a smart medical decision support system for the detection of heart disease, through several models such as LR, KNN, ANN, to classify individuals with cardiac diseases and individuals in a normal state. The proposal in [14] reached a 91.10% classification accuracy [14]. Other researchers used effective collection, pre-processing of data and data transformation methods to generate precise information for the training model for heart disease prediction. They used KNN for the missing data and Relief and LASSO for features extraction. The accuracy achieved is 99.05% [16]. Another study about heart disease prediction discussed in [17], researchers implemented the model prediction with 5 Active learning methods (MMC, Random, Adaptive, QUIRE, and AUDI). The experiments consist of using hyperparameter optimization by the grid search technique. [17]. Morshedul Bari Ant, at [18], focused on Alzheimer's disease. He applied ML algorithms to detect dementia. This system was created with the OASIS (Open Access series of imaging studies) dataset which trained by SVM, logistic regression, decision tree, and random forest models [18]. The study in [19] suggested ML approaches used with blood test data to forecast the likelihood of COVID-19-related death. A robust combination of five features predicts mortality with 96% accuracy. [19]. The main goal of the research in [20] was to explore the potential applications of big data analytics and machine learning-based techniques in diabetes. The proposed work achieves an accuracy of 83% [20]. Chronic Kidney Disease prediction was covered in the study [21], The researchers used the CKD dataset extracted from the UCI repository. Their major aim was the feature optimization. The developed model implemented 6 ML algorithms for training: ANN, C5.0, LR, LSVM, KNN, and random tree. The model's most significant accuracy in SMOTE with all features was 98.86% [21]. Shahadat Uddin compared the performances of the K-nearest neighbor (KNN) algorithm and its variants (Classic one, Adaptive, Locally adaptive, k-means clustering, Fuzzy, Mutual, Ensemble, Hassanat, and Generalised mean distance) for forecasting eight diseases [22]. A further interesting disease to explore in [36] about skin cancer, Researchers presented a machine learning (ML) classification-based automated image-based system for the recognition, extraction, processing, and categorization of the skin. The proposed method extracts the most valuable features from the skin images with an accuracy of 87% [36]. In [23], The researchers proposed ML and IoMT-based model to provide clinical decision support to reduce doctors' workload and decrease the death rate within the COVID-19 pandemic. The research described in [24] provided an ML model for a dataset of fundamental medical health that predicts what foods should be offered to particular patients according to their medical condition. The medical dataset consisted of 30 patient records with 13 features associated with various illnesses and 1000 items obtained from hospitals and the Internet. Andrei Velichk in [3] proposed a model for determining the presence of COVID-19 based on typical blood results. It applied 13 classifiers for examination and used the HGB approach for feature selection. The researchers used advanced Arduino computing and the IoMT cloud service [3]. The Table 3 summarizes researchers' studies of medical decision-making systems with machine learning and IoMT:

4.2   Deep Learning and IoMT-Based Diagnostic Applications

The study in [25] presented a DL architecture to detect tumor and non tumor tissue in histological images of CRC. the model achieved a nine-class accuracy of 94% [25]. An end-to-end deep learning system (DLS) was proposed by the researchers In [26] to predict Survival for patients with various forms of cancer [26]. Researchers have proposed a global architecture of pathological type identification of lung cancer during the early stage by CT images. From the experimental outcomes, VGG16-T with boost has an accuracy rate of 86.58% [27]. In [28] the research presented A Novel Method to detect COVID-19 through AI in Chest X-ray Images. The researchers proposed two approaches: The first one for COVID-19 classification and evaluation. The second one, for the feature extraction [28]. Other researchers proposed a new automated DL method for multiclass brain tumor classification [29]. In [30] researchers proposed a decision support system through physicians' knowledge that applied a fuzzy inference system (FIS). The reason for proposing such a system was that during the COVID-19 period, the datasets were not available. Therefore the solution was to benifit from the researchers' knowledge. Marwa EL-Geneedy in [31] developed a learning-based pipeline to recognize Alzheimer's multi-class disease with brain MRI images. The suggested approach provided both a local and a global categorization (i.e., normal vs. Mild Cognitive Impairment (MCI) vs. AD). The model reached 99.68% accuracy [31]. In [32], authors benefited from the advance of IoMT, and they proposed an IoMT-based fog calculation model to diagnose patients who suffered from type 2 diabetes [32]. Another study [33], consisted of a clinical decision support system with cloud-based IoMT for CKD prediction with DNN. The System gathers patient data through IoMT devices, and stored them in the cloud with their associated medical records from the UCI repository. The DNN classifier achieves a maximum classifier accuracy of 98.25% [33]. To extract precise characteristics from an MRI image, the researchers in [34] established a neural network model with a VGG16 feature extractor, with an accuracy of 90.40% for dataset 1 and 71.1% for dataset 2 [34]. The work [35] presented a new light-weight "Reduced-FireNet" deep learning based model for histopathology image self-classification [35].

Researchers [37] proposed a new approach to predict heart disease. They applied a pre-trained Deep Neural Network to extract feature, Principal Component Analysis (PCA) to reduce dimensionality, and Logistic Regression (LR) for prediction. The model suggested achieved an accuracy of 91.79% [37]. Roseline in [38] and other authors proposed an IoMT diagnosis system to identify breast cancer. The system consists of classifying the tissue into malignant and benign classes. The proposed model achieved a classification accuracy of 98.5% with CNN and 99.2% with ANN [38]. Oher research interest was about analyzing medical images using Deep Learning in real-time. The proposed system contains two parts: the first one aims to define the regions of interest (RoIs) where capillaries might exist, and the second part is to predict if the RoIs contain capillaries or not using CNN [39]. The study described in [40] focused on applying deep learning techniques to detect COVID-19. The proposed work used CNN architec-

Table 3. Machine learning and IoMT-based diagnostic applications

Ref.	Study aim	Year	Technologies used	Strength	Weakness	Data Type
[14]	a machine learning-based diagnostic approach to identifying heart disease	2018	Decision Tree Classifier, KNN, ANN, Naive Bayes, SVM, Logistic Regression	7 ML algorithms, 3 feature selection algorithms, the cross validation and 2 classifiers performance	Certain irrelevant features reduce the diagnostic system's performance and increase the calculation time	Multivariate (the Cleveland heart disease)
[32]	IoMT-based fog computing model to Diagnose Patients with Type 2 Diabetes	2019	Cloud Computing, fog computing, IoMT	The smart wearable served as both the measurement's tool and platform. The accuracy of the blood glucose measurement was assessed using 100-1000 large-scale simulations. They employed a hybrid procedure that combined the VIKOR method with type-2 neutrosphic technology.	a low accuracy prediction	Multivariate dataset
[16]	Prediction system for heart disease	2021	AdaBoost (AB), Decision Tree (DT), Gradient Boosting (GB), K-Nearest Neighbors (KNN), and Random Forest (RF), Relief, LASSO	Using 5 classifiers for prediction and 2 feature selection methods	The dependency on a specific Feature Selection technique and the missing values in the dataset	Multivariate (the Cleveland, Hungary, Switzerland, VA Long Beach Statlog heart disease

Table 3. (*continued*)

Ref.	Study aim	Year	Technologies used	Strength	Weakness	Data Type
[23]	ML and IoMT model for COVID-19 prediction in smart healthcare environments based on lab results.	2021	IoMT Random Forest, SVM	Combining the IoMT and ML (smart hospital environments), Various ML approaches were applied.	Large variations in the scales of the characteristics, excluded to just three ML models	multivariate dataset
[19]	A Clinical Decision Support System for Early COVID-19 Mortality Prediction Based on Machine Learning	2021	neural networks, logistic regression, XGBoost, random forests, SVM, and decision trees	The proposed model is based on blood tests data which make it more performance	They need more data from diverse resources	multivariate dataset
[18]	Predict Alzheimer's disease using ML Algorithms	2021	Support vector machine, logistic regression, decision tree, and random forest	Using multiple ML Algorithms to compare the results	Train small datasets	longitudinal Magnetic Resonance Imaging (MRI)
[21]	An examination of machine learning for the prediction of chronic kidney disease	2021	ANN, KNN, LSVM, C5.0, Chi-square, Logistic Regression, Filter method, Wrapper method, Embedded method	Using SMOTE class balancer	small dataset with many attributes	Multivariate
[3]	Routine Blood Values and Machine Learning Sensors for COVID-19 Disease Diagnosis for Internet of Things Application	2022	histogram-based gradient boosting, linear discriminant analysis, KNN, LSVM, NLSVM, passive-aggressive, multilayer perceptron, decision tree (DT)	Using the HGB approach for feature selection and 13 classifiers for disease prediction	The data represent mainly a unique institution the data set excludes patient comorbidity	Multivariate (SARS-CoV-2-RBV1)

Table 3. *(contniued)*

Ref.	Study aim	Year	Technologies used	Strength	Weakness	Data Type
[22]	Comparative performance analysis of KNN algorithm with its different variants for disease prediction	2022	Hassanat distance KNN (H-KNN), Generalised mean distance KNN, Mutual KNN, Ensemble approach KNN	applying the KNN algorithms with its variants	researchers didn't preprocess datasets a low model accuracy (high accuracy 86%) It ignores minority classes, which can affect its performance Inconsistent output for noisy data sets	Multivariate
[17]	Multi-Label Active Learning-Based Machine Learning Model for Heart Disease Prediction	2022	MMC, Random, Adaptive, QUIRE, and AUDI	using active learning methods for multi-label data	few labelled examples were used to train the initial classifier. Memorization still must to adapt	Multivariate
[24]	IoMT-Assisted Patient Diet Recommendation System Through ML Model	2020	RNN, GRU LSTM	combining IoMT with machine learning and deep learning to give a high prediction	low quality of IoMT used devices	Multivariate (Categorical and Numerical)

tures like VGG16, DeseNet121, MobileNet, NASNet, Xception, and EfficientNet [40]. The Table 4 summarizes researchers' studies of medical decision-making systems with Deep learning and IoMT: [h]

5 Discussion

This systematic review is interested in the recent proposed approaches of IoT and AI that are used in healthcare, while addressing their benefits and weaknesses. This review demonstrates that there is a significant increase in the number of the proposed works in this research area. The adoption of AI in medical field can speed up the process of tumor diagnosis without the need to the histological examination that can take a lot of time. This is due to the progress in CPU computing power, the efficiency of recent ML and DL algorithms and the accessibility

Table 4. Deep learning and IoMT-based diagnostic applications

Ref.	Study aim	Year	Technologies used	Strength	Weakness	Data Type
[25]	a model to recognize the different tissue types which are prevalent in the histological images of CRC (non-tumor tissue types).	2019	VGG16, GoogleNet, Alexnet, Squeezenet, Resnet50	The data is trained with CNN by transfer learning and reach a 9 classes	Before being used consistently in the clinic, this study needs to be prospectively validated. The technique of manually extracting tumor areas from whole slide histology images could be completely automated	Medical Images
[26]	An approach to provide significant prognostic information in several cancer types at specific pathological stages.	2022	CNN	Models were used to classify individual patients according to cancer stage (stage II). The CNN consisted of depth-separable convolution layers, identical to the MobileNet architecture, with adjusted size and number of layers to accelerate training and reduce the risk of overfitting	The approaches and outcomes can only be applied to datasets from TCGA (There are limited image)	histopathology images
[28]	A novel method CoVIRNet (COVID Inception-ResNet model) to identify the COVID-19 patients automatically using x-ray images.	2021	GAN, Deep Transfer Learning	The proposed deep training building blocks utilized different regularization techniques for minimizing overfitting due to the small COVID-19	The proposed technique's performance was tested on a small dataset of publically available chest radiographic images of different cases of pneumonia and COVID-19	X-ray Images

Table 4. *(continiued)*

Ref.	Study aim	Year	Technologies used	Strength	Weakness	Data Type
[27]	A CT image classification approach for advanced lung cancer based on the VGG16 convolutional neural network	2020	VGG16	The researchers built their own dataset containing 2219 CT-images and trained it with VGG16 and applied the boosting strategy to aggregate multiple classification outcomes to enhance the performance of classification methods.	More contextual tumor information, such as connections to surrounding blood vessels, and patient information, such as a medical history report, can be merged	CT images
[29]	An innovative automated deep learning approach for detecting multiple-class brain tumors.	2021	Densenet201 CNN	A pre-trained model (the Densenet201) for multiclass brain tumor classification. The use of MGA and Entropy-Kurtosis-based methods to choose the best features. The fusion of the optimal features	the reduction of certain important features	MRI Images (BRATS2018 and BRATS2019)
[30]	A decision support system based on doctor knowledge to decrease stress in the community, to disrupt the chain of propagation of COVID-19	2020	Fuzzy inference system (FIS)	The system based on expert's knowledge	The system is based on only 3 criteria: fever, fatigue and dry cough	Textual Dataset

Table 4. *(contniued)*

Ref.	Study aim	Year	Technologies used	Strength	Weakness	Data Type
[33]	An ioMT with cloud-based clinical decision support system for CKD prediction and adherence with severity level	2019	DNN IoMT	they used a feature selection method based on Particle Swarm Optimization (PSO)	poor CDS security	Multivariate
[34]	A deep learning based convolutional neural network model with VGG16 feature extractor to identify Alzheimer Disease through MRI scans	2022	VGG16 CNN	Using the VGG16 for feature extraction and use two datasets for evaluation	Small datasets	MRI Images
[31]	An MRI-based deep learning method for precise Alzheimer's disease identification	2022	CNN, Transfer Learning	Employing the augmentation approach to produce new images (to solve the problems of data lack)	prediction for only one stage of Alzheimer's disease	MRI Scans
[35]	A noval Lightweight Deep Learning Model for Classifying Histopatholog-ical Images for IoMT	2021	CNN, SqueezeNet	Using the IoMT for disease identification in real time with deep learning model. A new model with small size to capture medical images and process information protect the medical images of patient	IoMT Devices have limited computational power and memory space	Multivariate dataset

Table 4. *(continued)*

Ref.	Study aim	Year	Technologies used	Strength	Weakness	Data Type
[36]	Classifier for predict 6 skin disease	2023	CNN SVM	Classifier for predict 6 skin disease	low accuracy	CT images
[24]	IoMT-Assisted Patient Diet Recommendation System Through Machine Learning Model	2020	RNN, GRU and LSTM	combining IoMT with machine learning and deep learning to give a high prediction	IoMT Devices have limited computational power and memory space	Multivariate dataset
[37]	A heart disease prediction with deep learning	2023	PCA Logistic Regression DNN	A number of DNN structures for prediction evaluation. Apply the PCA	Researchers didn't use Feature Extraction Algorithms	multivariate dataset
[38]	IoMT and Deep learning based model for breast cancer prediction	2022	SVM, ANN, CNN, PSO	Employing the PSO as feature selection approach Integrate the medical devices (IoMT)	Researchers need large data	multivariate dataset
[39]	A Software Design Pattern for Deep Learning-Based Real-Time Medical Image Analysis	2022	Cloud, DNN, Hadoop Distributed File System	detect the callipers in heart disease and COVID-19 in real time Use 3 metrics to evaluate the proposed model	a low hardware capacities	Medical Images
[40]	CNN architecture to detect COVID-19 through lung CT Scan.	2022	VGG16, DeseNet121, MobileNet, NASNet, Xception, and EfficientNet	Using Multiple pre-trained models for predicting Using Lung CT scan	A small datastes and few number of epochs.	CT scan Images

of huge amounts of data (big data) collected from electronic health monitors and medical health records. The presented work in this paper covered the involvement of machine and DL models along with IoMT in the diagnosis of cancer, diabetes, chronic diseases, heart, Alzheimer, and COVID. The machine learning models used include random forest classifier, logistic regression, decision tree, K-nearest neighbor (KNN), and support vector machines (SVM). Moreover, the deep learning models used include convolutional neural networks (CNN) have often been used for disease diagnosis and pre-trained models: VGG16, GoogleNet, Alexnet, Squeezenet, Resnet50. Despite the significant advantages of AI and IoT-based techniques in diagnosing diseases, many challenges remain to be overcome such as:

1. The small size of the pre-processed datasets that can be used to evaluate the proposed approaches efficiency.
2. The lack of efficient feature selection techniques to clean medical data while generating high precision in disease prediction.
3. The problems of data security during the collection phase.
4. The absence of models validation phase by medical experts.

6   Conclusion and Future Perspectives

AI is a broad field of approaches that combines data mining and analytics, machine learning and deep learning, data collection, and pattern recognition that is regularly evolving and growing to fit the criteria needed of the Healthcare sector and its Patients. This article explores the potential impact of technologies such as IoMT and AI in healthcare and the clinical support system through a systematic literature review of 27 published papers. To clarify the performance of AI and IoMT in clinical support systems, This current study was divided into three sections covering applications for machine learning- and deep learning-based diagnostics, and IoMT-based applications. The main conclusions build across this paper state that the approaches of AI and IoMT in the clinical support system for disease detection, prediction, and patient monitoring are an avoidable path to migrate toward precision medicine. However, problems like lack of data, ethical and data security, and non-practical accuracy represent the major research challenges that need to be addressed. Therefore, in future work, we aim to elaborate and introduce technical and research solutions for aiding intelligent medical support systems to provide secure and robust models. Due to the continuous progress in artificial intelligence and IoMT fields, there are many opportunities for extending the research presented in this work. Although our SLR has highlighted the potential of AI and IoMT technologies to predict and prevent the onset of various diseases, it still needs improvement to be suitable for the medical field. In our future works, we will focus on the potential of innovative deep learning and machine learning methods. Furthermore, to improve the predictive performance of these models, researchers need to integrate other data sources such as genetic and environmental factors. To determine the most effective fusion among data sources and modeling approaches for various illness types, additional investigation is also needed. Real-time illness forecasting systems which can be implemented to support medical decision making is another research area in our further studies.

References

1. Merabet, A., Ferradji, M.A.: Smart virtual environment to support collaborative medical diagnosis. In: 2022 4th International Conference on Pattern Analysis and Intelligent Systems (PAIS), pp. 1–6. IEEE, October 2022

2. Afrash, M.R., Erfanniya, L., Amraei, M., Mehrabi, N., Jelvay, S., Shanbehzadeh, M.: Machine learning-based clinical decision support system for automatic diagnosis of COVID-19 based on the routine blood test. J. Biostat. Epidemiol. 8(1), 77–89 (2022)

3. Velichko, A., Huyut, M.T., Belyaev, M., Izotov, Y., Korzun, D.: Machine learning sensors for diagnosis of COVID-19 disease using routine blood values for internet of things application. Sensors 22(20), 7886 (2022)

4. Ibrahim, I., Abdulazeez, A.: The role of machine learning algorithms for diagnosing diseases. J. Appl. Sci. Technol. Trends 2(01), 10–19 (2021)

5. Kashani, M.H., Madanipour, M., Nikravan, M., Asghari, P., Mahdipour, E.: A systematic review of IoT in healthcare: applications, techniques, and trends. J. Netw. Comput. Appl. 192, 103164 (2021)

6. Mahesh, B.: Machine learning algorithms-a review. Int. J. Sci. Res. (IJSR). [Internet] 9, 381–386 (2020)

7. Feng, Y., Wang, Y., Zeng, C., Mao, H.: Artificial intelligence and machine learning in chronic airway diseases: focus on asthma and chronic obstructive pulmonary disease. Int. J. Med. Sci. 18(13), 2871 (2021)

8. Tiwari, D., Bhati, B.S., Al-Turjman, F., Nagpal, B.: Pandemic coronavirus disease (COVID-19): world effects analysis and prediction using machine-learning techniques. Expert. Syst. 39(3), e12714 (2022)

9. Nusinovici, S., et al.:Logistic regression was as good as machine learning for predicting major chronic diseases. J. Clin. Epidemiol. 122, 56–69 (2020)

10. Shamshirband, S., Fathi, M., Dehzangi, A., Chronopoulos, A.T., Alinejad-Rokny, H.: A review on deep learning approaches in healthcare systems: taxonomies, challenges, and open issues. J. Biomed. Inform. 113, 103627 (2021)

11. Suganyadevi, S., Seethalakshmi, V., Balasamy, K.: A review on deep learning in medical image analysis. Int. J. Multimedia Inf. Retrieval 11(1), 19–38 (2022)

12. Gatouillat, A., Badr, Y., Massot, B., Sejdić, E.: Internet of medical things: a review of recent contributions dealing with cyber-physical systems in medicine. IEEE Internet Things J. 5(5), 3810–3822 (2018)

13. Manickam, P., Mariappan, S.A., Murugesan, S.M., Hansda, S., Kaushik, A., Shinde, R., Thipperudraswamy, S.P.: Artificial intelligence (AI) and internet of medical things (IoMT) assisted biomedical systems for intelligent healthcare. Biosensors 12(8), 562 (2022)

14. Haq, A.U., Li, J.P., Memon, M.H., Nazir, S., Sun, R.: A hybrid intelligent system framework for the prediction of heart disease using machine learning algorithms. Mob. Inf. Syst. (2018)

15. Ahsan, M.M., Luna, S.A., Siddique, Z.: Machine-learning-based disease diagnosis: a comprehensive review. In: Healthcare, vol. 10, no. 3, p. 541. MDPI, March 2022

16. Ghosh, P., et al.:Efficient prediction of cardiovascular disease using machine learning algorithms with relief and LASSO feature selection techniques. IEEE Access 9, 19304–19326 (2021)

17. El-Hasnony, I.M., Elzeki, O.M., Alshehri, A., Salem, H.: Multi-label active learning-based machine learning model for heart disease prediction. Sensors 22(3), 1184 (2022)

18. Bari Antor, M., et al.: A comparative analysis of machine learning algorithms to predict Alzheimer's disease. J. Healthc. Eng. (2021)
19. Karthikeyan, A., Garg, A., Vinod, P.K., Priyakumar, U.D.: Machine learning based clinical decision support system for early COVID-19 mortality prediction. Front. Publ. Health **9**, 626697 (2021)
20. Krishnamoorthi, R., et al.: A novel diabetes healthcare disease prediction framework using machine learning techniques. J. Healthc. Eng. (2022)
21. Chittora, P., et al.: Prediction of chronic kidney disease-a machine learning perspective. IEEE Access **9**, 17312–17334 (2021)
22. Uddin, S., Haque, I., Lu, H., Moni, M.A., Gide, E.: Comparative performance analysis of K-nearest neighbour (KNN) algorithm and its different variants for disease prediction. Sci. Rep. **12**(1), 1–11 (2022)
23. Abdulkareem, K.H., et al.: Realizing an effective COVID-19 diagnosis system based on machine learning and IOT in smart hospital environment. IEEE Internet Things J. **8**(21), 15919–15928 (2021)
24. Iwendi, C., Khan, S., Anajemba, J.H., Bashir, A.K., Noor, F.: Realizing an efficient IoMT-assisted patient diet recommendation system through machine learning model. IEEE Access **8**, 28462–28474 (2020)
25. Kather, J.N., et al.: Predicting survival from colorectal cancer histology slides using deep learning: a retrospective multicenter study. PLoS Med. **16**(1), e1002730 (2019)
26. Wulczyn, E., et al.: Deep learning-based survival prediction for multiple cancer types using histopathology images. PloS ONE **15**(6), e0233678 (2020)
27. Pang, S., et al.: VGG16-T: a novel deep convolutional neural network with boosting to identify pathological type of lung cancer in early stage by CT images. Int. J. Comput. Intell. Syst. **13**(1), 771 (2020)
28. Almalki, Y.E., et al.: A novel method for COVID-19 diagnosis using artificial intelligence in chest X-ray images. In: Healthcare, vol. 9, no. 5, p. 522. Multidisciplinary Digital Publishing Institute, May 2021
29. Sharif, M.I., Khan, M.A., Alhussein, M., Aurangzeb, K., Raza, M.: A decision support system for multimodal brain tumor classification using deep learning. Complex Intell. Syst. **8**(4), 3007–3020 (2022)
30. Govindan, K., Mina, H., Alavi, B.: A decision support system for demand management in healthcare supply chains considering the epidemic outbreaks: a case study of coronavirus disease 2019 (COVID-19). Transport. Res. Part E Logist. Transport. Rev. **138**, 101967 (2020)
31. Marwa, E.G., Moustafa, H.E.D., Khalifa, F., Khater, H., AbdElhalim, E.: An MRI-based deep learning approach for accurate detection of Alzheimer's disease. Alex. Eng. J. **63**, 211–221 (2023)
32. Abdel-Basset, M., Manogaran, G., Gamal, A., Chang, V.: A novel intelligent medical decision support model based on soft computing and IoT. IEEE Internet Things J. **7**(5), 4160–4170 (2019)
33. Lakshmanaprabu, S.K., Mohanty, S.N., Krishnamoorthy, S., Uthayakumar, J., Shankar, K.: Online clinical decision support system using optimal deep neural networks. Appl. Soft Comput. **81**, 105487 (2019)
34. Sharma, S., Guleria, K., Tiwari, S., Kumar, S.: A deep learning based convolutional neural network model with VGG16 feature extractor for the detection of Alzheimer disease using MRI scans. Meas. Sens. **24**, 100506 (2022)
35. Datta Gupta, K., Sharma, D.K., Ahmed, S., Gupta, H., Gupta, D., Hsu, C.H.: A novel lightweight deep learning-based histopathological image classification model for IoMT. Neural Process. Lett. 1–24 (2021)

36. Yu, Z., Wang, K., Wan, Z., Xie, S., Lv, Z.: Popular deep learning algorithms for disease prediction: a review. Cluster Comput. 1–21 (2022)

37. Hassan, D., Hussein, H.I., Hassan, M.M.: Heart disease prediction based on pre-trained deep neural networks combined with principal component analysis. Biomed. Sig. Process. Control **79**, 104019 (2023)

38. Ogundokun, R.O., Misra, S., Douglas, M., Damaševičius, R., Maskeliūnas, R.: Medical internet-of-things based breast cancer diagnosis using hyperparameter-optimized neural networks. Future Internet **14**(5), 153 (2022)

39. Abdou, M.A.H., Ferreira, P., Jul, E., Truong, T.T.: Capillaryx: a software design pattern for analyzing medical images in real-time using deep learning. arXiv preprint arXiv:2204.08462 (2022)

40. Kogilavani, S.V., et al.: COVID-19 detection based on lung CT scan using deep learning techniques. Comput. Math. Methods Med. (2022)

Feature Embedding Representation for Unsupervised Speaker Diarization in Telephone Calls

Meriem Hamouda[1,2] and Halima Bahi[1,2(✉)]

[1] LISCO Laboratory, Badji Mokhtar University, Annaba, Algeria
`meriem.hamouda@univ-annaba.org, halima.bahi@univ-annaba.dz`
[2] Computer Science Department, Badji Mokhtar University, Annaba, Algeria

Abstract. Speaker diarization aims to segment an audio recording, where different speakers are involved, into speech segments based on the speaker's identity. This study proposes a feature embedding representation for unsupervised speaker diarization in case of telephone conversations. First, the speech waveform is segmented into speech and non-speech segments using a voice activity detection algorithm. Then, a deep autoencoder performs feature embedding representation of the speech segments which are fed to the K-means algorithm to perform the clustering. Indeed, in the case of telephone calls speech data is often unequally shared between the two speakers, and an accurate feature representation helps to overcome limitations related to the varying width across clusters of the K-means algorithm. Experiments carried out on the CALLHOME dataset show the positive impact of the embedding representation in terms of Diarization Error Rate (DER) with an improvement of about 56%.

Keywords: Speaker Diarization · Unsupervised Speech Segmentation · Clustering · Voice Activity Detection (VAD) · Feature embedding · Autoencoder · K-Means

1 Introduction

Technological advances in favor of communication and the opening of social networks to spoken communication have made a large amount of speech data available. Taking advantage of these data involved various speech processing applications. Speech diarization is one of these applications that aims to segment an audio recording, where different speakers are involved, into speech segments based on the speakers' identities [1–4].

Given a speech segment from a speech flow, speaker diarization consists of identifying the person who speaks in this segment, and when the segment occurs in the flow. The need to perform diarization has become very useful in remote meetings, telecommuting, medical diagnostics, and many other applications sparked by the widespread use of smart devices. In many situations, the identity of the speakers is unknown to segment the signal according to the given identities, herein, approaches of unsupervised speaker diarization are required.

A. Bennour et al. (Eds.): ISPR 2023, CCIS 1940, pp. 207–215, 2024.
https://doi.org/10.1007/978-3-031-46335-8_16

Broadly, a typical speaker diarization framework can be summarized into pre-processing, feature extraction, segmentation, clustering, and case-dependent post-processing [4, 5].

First, the voice activity detection module detects speech and non-speech regions from the considered signal, then it removes the non-speech regions. Thus, it contributes to reducing the number of required computations in the further stages. Many studies dealt with voice activity detection to improve the performance of the speaker diarization algorithm. Bhuyan et al. (2022) used both the energy and the Bayesian Information Criterion (BIC) to perform the segmentation [1]. Their experiments carried on short audio conversations, collected from zoom meetings, YouTube and podcasts, showed a reduction of the false positive rate. The authors claimed that their approach can be applied in real-time speaker diarization, particularly, on low-power embedded devices [1]. Jiang and Liu (2020) used the spectrum centroid, and the local maxima of the statistical feature sequence histogram to set the threshold for the segmentation [6].

Then, as the speech signal is highly redundant, the feature extraction stage is crucial; it serves to extract pertinent characteristics from the speech signal. In speaker diarization applications, short-term spectral features such as Mel-Frequency Cepstral Coefficients (MFCC) are state-of-the-art. These features encapsulate average information about the vocal tract characteristics of a speaker, however, they are also sensitive to elements unrelated to the speaker's identity. To eleviate the limitation related to the use of MFCC features, Yella et al., (2014) combined the MFCC with features extracted from an artificial neural network (ANN) in a dual-stream mode [7]. The results show a significant improvement in the diarization error of about 14% in favor of the combination (ANN+MFCC) [7].

After that, the detected speech regions are split into short segments to build appropriate speaker models. Herein, the most used modeling approaches are the Hidden Markov Models (HMM), Gaussian Mixture Models (GMM), and Support Vector Machines (SVM) [8]. Finally, the obtained segments are clustered according to the speakers' identities using different clustering methods, such as K-means, Agglomerative Hierarchical Clustering (AHC), spectral clustering, and the integral path clustering [9].

Many works in the literature have proposed clustering-based speaker diarization methods. In [10], the authors combined self-supervision-based metric learning with a graph-based clustering algorithm. In [11], an analysis of the speaker diarization pipeline that uses embeddings of d-vectors based on Long Short-Term Memory (LSTM) on various clustering techniques is performed, the spectral clustering performed well on the used dataset and resulted in the lowest error rate. In [12], a framework is proposed, it solicits a human in the loop to correct the clustering by answering simple questions. The authors in [13] proposed the use of a two-speaker end-to-end diarization method as a post-processing of the results obtained by a clustering method.

This paper is interested in speaker diarization in the case of telephone conversations. In such cases, the speakers are often unknown, thus the use of an unsupervised segmentation method is required. In particular, as in telephone conversations, only two speakers are involved, the use of K-means seems a valuable suggestion as the number of clusters is previously known. However, one of the K-means clustering algorithm drawbacks is its difficulty to handle the varying width across clusters, which is a situation often encountered in telephone conversations. To overcome this limitation, we suggest modeling the incoming speech using the GMM models. Meanwhile, recent advances made in deep learning sparkled research in feature extraction. Thus to better capture speakers' characteristics, we intend to use a deep autoencoder (DAE) to perform a feature embedding representation at the acoustic level. The experimental results on the CALLHOME dataset [14] reveal that the proposed method achieved significant improvement.

The rest of the paper is organized into sections as follows. Section 2 describes in depth the proposed framework. The results obtained on the CALLHOME dataset are presented and discussed, in Sect. 3. Finally, a conclusion is drawn.

2   Unsupervised Speaker Diarization Based on Feature Embeddings Representation and K-Means Algorithm

In telephone calls, it is expected to have two protagonists, thus the clustering algorithm gathers the speech segments into two clusters standing for the two speakers.

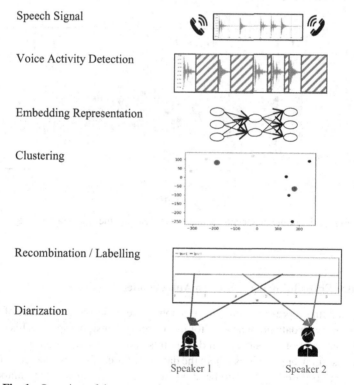

Fig. 1. Overview of the proposed unsupervised speaker diarization system

The proposed speaker diarization system is illustrated in Fig. 1. It includes a voice activity detection module based on short-term energy, a feature extraction module that extracts Mel Frequency Cepstral Coefficients (MFCCs) which are fed to a deep autoencoder to provide the embedding representation, and a clustering module. The results of the clustering allow us to segment the incoming speech, herein the telephone conversation, and to label the obtained segments according to the various clusters standing for the speakers' identities.

2.1 Voice Activity Detection (VAD)

Voice activity detection aims to segment the conversation into speech and non-speech segments. For that purpose, many methods exist. In particular, short-term energy is one of the most used features in automatic speech and speaker segmentation as well. The signal is split into frames using windowing, and a threshold is set. If the energy within a given segment exceeds the predefined threshold, the segment is considered speech otherwise, it is considered non-speech. In the case of telephone calls, the non-speech segments tend to be silence or noisy background, thus, they are discarded from further processing. Figure 2 illustrates the automatic segmentation of the conversation #4520 from the CALLHOME dataset.

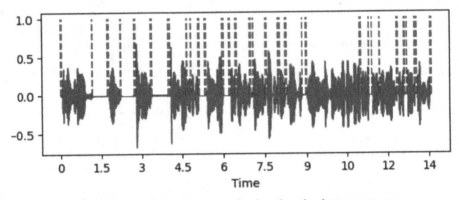

Fig. 2. Example of speech segmentation based on the short-term energy

2.2 Feature Embedding using a Deep Autoencoder

Feature embedding representation aims to capture the underlying structure of the data, by representing each data point as a vector in a lower-dimensional space. This brings out similarities and relationships between data points and allows their comparison. This is particularly useful in applications where the data are of great dimensionality as in speech signals. In the proposed feature embedding method, we employ a deep autoencoder-based unsupervised method to learn the new features from the hand-crafted MFCC features.

DAE is a feedforward neural network with many hidden layers. "A deep autoencoder is trained to reconstruct the input X on the output layer Y through one (or more) hidden layer H" [15, p. 13]. Given a pattern X, the encoding layers reduce X into a smaller dimension, then, the decoder layers attempt to reconstruct the pattern Y which is a version as close as possible to X [16]. In this work, the feature embedding representation of a speech segment provides an acoustic vector where adjacent vectors are considered. The DAE is trained over three context lengths of the target frame. Thus, the deep representation of the target frame is learned by considering the surrounding context; in this case, one frame from each side. The motivation is that speaker change usually does not occur in three successive frames, which allows the DAE to capture the context of the current frame considering the speaker's particularities (Fig. 3).

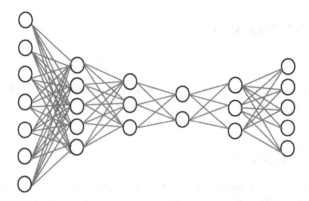

Fig. 3. Deep autoencoder for features embedding representation

In this work, 20-MFCC coefficients are extracted from each frame. Then, 3 × 20 inputs are fed to the DAE; inputs represent the coefficients of the target frame as well as those of the surrounding two ones. The DAE outputs a vector of 20 values standing for the middle frame (the target) feature embedding representation.

2.3 GMM Modeling and Clustering

Clustering looks for a typology, or segmentation, where a distribution of individuals into homogeneous classes is performed. In the case of telephone conversations, the speakers are grouped into two clusters. The results are easy to interpret, as the clusters stand for speakers' identities.

For this work, we choose the K-means algorithm to perform the clustering since the number of clusters is previously known. This mode of clustering works great when dealing with spherical clusters, however, K-means is known to perform worse when the clusters have varying widths, a situation often encountered in telephone conversations. To overcome this limitation, and alongside the feature embedding representation, we suggest modeling the speech segments using a 32-mixture GMM.

2.4 Segmentation and Labeling

Once the clustering is performed, speaker hypothesis values are determined at the segment level; then, segments are converted to frame level and labeled based on the start time and the end time of the segments. Table 1 shows an excerpt of an illustrative diarization.

Table 1. Example of the diarization result

#Segment	Speaker Label	Start Time	End Time
0	Speaker 1	0.44	3.62
1	Speaker 0	3.62	15.52
2	Speaker 1	15.52	19.36
3	Speaker 0	19.36	35.56

3 Results and Discussion

3.1 Dataset

The experiment was carried out on telephone conversations from the English part of CALLHOME corpus [14], where only two speakers conversations were selected as the clustering is limited to two clusters in our case.

3.2 Evaluation Metrics

To evaluate the proposed speaker diarization algorithm, we consider the Diarization Error Rate (DER). DER is the standard speaker diarization performance metric, it takes into account both the false alarm and missed speaker error rates [10]. DER is the overall percentage of the reference speaker time that is not accurately attributed to the speaker, it is defined as the sum of three errors: false alarm of speech (FA), missed detection of speech (Miss), and confusion between speaker labels (Err) [17]:

$$DER = \frac{FA + Miss + Err}{totalDuration\ of\ Time} \tag{1}$$

Besides, the DER, the precision and recall measures, traditionally used to assess classification algorithms, are also computed.

3.3 Experimental Results

To assess the impact of the feature embedding on the diarization performance, we compared the results of the proposed method with those of the K-means clustering with and without the GMM modelling based on the MFCC. The results are shown in Table 2.

In this work, the feature embedding representation of the speech signal is compared to hand-crafted features for unsupervised speaker diarization task in the context of telephone conversations. The clustering is performed by the K-means.

Table 2 shows that the feature embedding positively impacts the three scores. Table 2 also shows the beneficial effect of the GMM on the K-means where the DER is drastically reduced.

Table 2. Results of the proposed method compared to the K-means and GMM/K-means

	MFCC +K-means	MFCC +GMM/K-Means	Deep feature embedding +GMM/K-Means
Precision(%)	55.83	87.30	95.06
Recall(%)	57.85	90.42	94.86
DER(%)	63.78	15.32	6.94

3.4 Illustrative Example

As an illustration of the speaker diarization proposed method, we consider the recording of the telephone conversation #4520. Figure 4 reports the timesheets of the various segmentation and labeling methods, respectively, MFCC+K-means, MFCC+GMM/K-means, and deep embeddings representation+GMM/K-means. The segmentation and labeling manually performed and shown in Fig. 4a are considered as the reference.

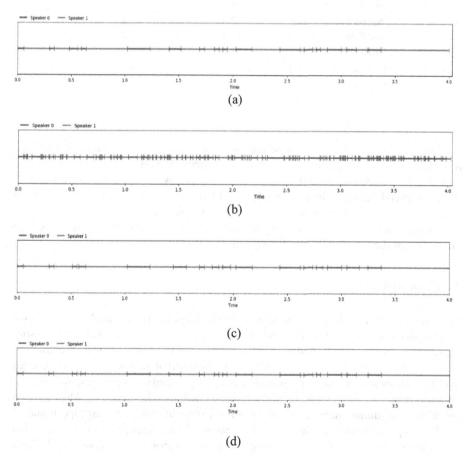

Fig. 4. Diarization results as a timesheet using: (a) Manual diarization, (b) MFCC+K-means (c) MFCC+GMM/K-means, (d) Deep representation+GMM/K-means

Compared to the manual diarization, the K-means shows a high rate of false identification (Fig. 4a and Fig. 4b). A significant improvement is observed with the GMM/K-Means diarization (Fig. 4c). However, the best results are observed with the deep representation+GMM/K-means algorithm (Fig. 4d) with a low error rate.

To better illustrate the impact of the proposed method, Table 3 reports the results of four other recordings, in terms of precision, recall, and diarization error rate.

Table 3. Illustrative example

#Recording	Measures	MFCC +K-means	MFCC +GMM/K-Means	Deep feature embedding +GMM/K-Means
#4390	Precision(%)	50.4	98	100
	Recall(%)	57.8	98	97.8
	DER(%)	59.4	18.4	12.4
#4547	Precision(%)	66.8	91.5	91.5
	Recall(%)	65.5	91.5	91.5
	DER(%)	70.3	8.5	8.5
#4484	Precision(%)	44.8	77.4	92.8
	Recall(%)	49.6	77.4	92.8
	DER(%)	56.8	22.6	7.2
#4854	Precision(%)	77.9	91.5	95.2
	Recall(%)	75.6	91.5	97.3
	DER(%)	50.9	25.4	4.9

Table 3 shows the disparities of the performance measures between the various recordings. Overall, it can be seen that the two methods MFCC+GMM/K-means and deep feature embedding+GMM/K-means give similar results when the conversation is rather balanced between the two speakers (#4390 and #4547), otherwise the method with feature embedding performs better (#4484 and #4854).

Experiments have also shown that the best results are found in the recordings where the gender of the two speakers is different, as in #4854.

4 Conclusion

Speaker diarization is a challenging task involved in many applications. In this work, we propose an unsupervised speaker diarization algorithm for telephone convesations using the Gaussian mixture model and K-means clustering. In this work, the feature extraction stage is investigated to improve the results on the speaker diarization. Thus, the traditional MFCC features are compared to a proposed deep feature embedding representation. The obtained results show the benefits of the proposed features. In particular, when the two speakers have similar characteristics or when the speaking times are disproportionate.

The proposed method has also shown potential in handling segments with an overlap of speakers. However, additional tunings have to be carried to improve the obtained results. Indeed, unsupervised speaker diarization remains a challenging activity, where the use of deep learning algorithms is beneficial.

References

1. Bhuyan, A. K., Dutta H., Biswas, S.: Unsupervised quasi-silence based speech segmentation for speaker diarization. In: 9th International Conference on Sciences of Electronics, Technologies of Information and Telecommunications (SETIT), pp. 170–175 (2022). https://doi.org/10.1109/SETIT54465.2022.9875932
2. Tranter, S.E., Reynolds, D.A.: An overview of automatic speaker diarization systems. IEEE Trans. Audio Speech Lang. Process. **14**(5), 1557–1565 (2006)
3. Anguera, X., Bozonnet, S., Evans, N., Fredouille, C., Friedland, G., Vinyals, O.: Speaker diarization: a review of recent research. IEEE Trans. Audio Speech Lang. Process. **20**(2), 356–370 (2012)
4. Moattar, M.H., Homayounpour, M.M.: A review on speaker diarization systems and approaches. Speech Commun. **54**(10), 1065–1103 (2012)
5. Teimoori, F., Razzazi, F.: Unsupervised help-trained LS-SVR-based segmentation in speaker diarization system. Multimedia Tools Appl. **78**(9), 11743–11777 (2019)
6. Jiang, N., Liu, T.: An improved speech segmentation and clustering algorithm based on SOM and k-means. Mathematical Problems in Engineering (2020)
7. Yella, S. H., Stolcke, A.: Slaney, M.: Artificial neural network features for speaker diarization. In 2014 IEEE Spoken Language Technology Workshop (SLT), 402–406 (2014)
8. Teimoori, F., Razzazi, F.: Incomplete-data-driven speaker segmentation for diarization application; a help-training approach. Circuits Syst. Signal Process. **38**(6), 2489–2522 (2019)
9. Wang, W., Lin, Q., Cai, D., Li, M.: Similarity measurement of segment-level speaker embeddings in speaker diarization. IEEE/ACM Trans. Audio Speech Lang. Process. **30**, 2645–2658 (2022)
10. Singh, P., Ganapathy, S.: Self-supervised representation learning with path integral clustering for speaker diarization. IEEE/ACM Trans. Audio Speech Lang. Process. **29**, 1639–1649 (2021)
11. Gupta, A., Purwar, A.: Analysis of clustering algorithms for Speaker Diarization using LSTM. In: 1st International Conference on Informatics (ICI), pp. 19–24 (2022)
12. Shamsi, M., Barrault, L., Meignier, S., Larcher, A.: Active correction for speaker diarization with human in the loop. In: Iberspeech (2021)
13. Horiguchi, S., Garcia, P., Fujita, Y., Watanabe, S., Nagamatsu, K.: End-to-end speaker diarization as post-processing. In: International Conference on Acoustics, Speech and Signal Processing (ICASSP), pp. 7188–7192 (2021)
14. Canavan, A., Graff, D., Zipperlen, G.: CALLHOME American English Speech LDC97S42. LDC Catalog, Philadelphia: Linguistic Data Consortium (1997)
15. Frihia, H., Bahi, H.: One-class training for intrusion detection. In: Proceedings of the 1st International Conference on Intelligent Systems and Pattern Recognition (ISPR), pp. 12–16 (2020)
16. Dendani, B., Bahi, H., Sari, T.: Speech enhancement based on deep AutoEncoder for remote Arabic speech recognition. In: Image and Signal Processing: 9th International Conference, ICISP 2020, Marrakesh, Morocco, pp. 221–229 (2020)
17. Park, T.J., Kanda, N., Dimitriadis, D., Han, K.J., Watanabe, S., Narayanan, S.: A review of speaker diarization: recent advances with deep learning. Comput. Speech Lang. **72**, 101317 (2022)

Using Blockchain Technology for Ranking the Common Core Cycles' Students in Algerian Universities

Noura Zeroual[✉], Lamia Mahnane, and Mohamed Hafidi

LRS Laboratory, Computer Science Department, Badji Mokhtar University, Annaba, Algeria
noura.zeroual2017@gmail.com

Abstract. Higher education is regarded as the final phase of the educational system and one of the pillars for the advancement of society. It establishes development strategies for the improvement of society and scientifically qualifies students to work in a variety of societal sectors. However, the North African universities in general and Algerian universities in particular continue to classify their students according to traditional systems in the many universities disciplines. In order to rank its students in the pre-specialization phases, Algerian universities continue to use the LMD system (License - Master - Doctorate). The subjects of the pre-specialty and the outcomes of the specialty of the previous promotions are not taken into account in this method, which is based on the average of classification as an orientation criterion. The fact that the current educational systems in Algeria rely mostly on central storage devices managed by a single authority also makes them more susceptible to the loss of this data in the event that a central server fails or is compromised by hacking operations. The goals of this article are to create a system based on blockchain technology for the automatic classification of students from the second year of the law department at the University of Skikda, Algeria, to the third year using the academy's history and the results of previous promotions.

Keywords: Blockchain · Student ranking · Higher education · LMD · SVM classifier model

1 Introduction

In recent years, the world has witnessed many developments in all areas of life, such as industry, healthcare, economy especially higher education which is the top of the educational pyramid and one of the most important pillars on which nations and societies are based, as it contributes to building up human resources for the advancement and development of society. Higher education in Algeria has undergone many changes, including: teaching practices are changing under the influence of social changes, increased student mobility, etc. However, despite the realization of the need to introduce and use different technologies to keep pace with various developments in the field of higher education in developed countries, the administrations of North African universities in general and

A. Bennour et al. (Eds.): ISPR 2023, CCIS 1940, pp. 216–227, 2024.
https://doi.org/10.1007/978-3-031-46335-8_17

Algerian in particular still depend on the LMD system (Bachelor - master - doctorate) in the orientation of students from the second to the third university year. The LMD system is based on the average of classification as orientation criterion without taking into account the relation between the subjects of the second year and the results of the third year of the preceding promotions to classify and orient students [1]. Moreover, this method does not reflect the fact that a particular student has actually acquired these skills and achieved these results. Consequently, this method oblige students to accept specialties which are not of their choice and do not correspond to their ambitions and their future aspirations, which can lead to the failure of these students in the specialty they have been obliged to study.

The current education systems in Algeria also depend primarily on having all of their data stored centrally at a single central server that is supervised by a single authority, which makes them more susceptible to losing that data in the event that the central server fails or is targeted by hacking attempts [2]. These centralized solutions are also more vulnerable to fraud and data hacking. However, these systems have a reputation for being slow to reply to user requests, particularly those that are time-sensitive or depend on multiple users accessing the server at once. As a result, these systems have a number of issues, including the fact that they can only be used with a central authority's approval and lack confidence due to their susceptibility to fraud and hacking. Blockchain technology is one of the most popular technologies in the last years. Its first practical use was in 2008, as a digital layer to record simple peer-to-peer transactions of the crypto-currency bitcoin [3]. The main objective of its use was to eliminate any intervention by third-party intermediaries, and to allow users to carry out their transactions directly. Subsequently, the unique characteristics of this technology such as decentralization, immutability, security, reliability and loyalty [4] have attracted the attention of researchers and practitioners in other fields as well as that of the education. The explanation of this technology can be done in the form of a permanent ledger that keeps track of all data transactions that have occurred in chronological order [3]. The blockchain consists of a chain of blocks. Each block contains data or records of interest (transactions) and is linked to the previous block in the chain by a hash pointer data structure [5]. Blockchain can be seen as a special form of decentralized databases designed to allow multiple blockchain network participants to achieve an "agreement" on changes in the state of shared data without the need for a trusted central authority [5]. Therefore, blockchain technology can be a great benefit for higher education which can be used to record the various information about their students' learning abilities and progress in a decentralized, transparent and more secure way; thus to certify the performances on each subject or success' level of the students and share them on a network of other universities, companies and related governmental entities.

The purpose of this research is to assist students in selecting an appropriate course of study based on their abilities and the relationship between their achievements and their future success in a specific discipline. To accomplish this, we designed a system based on blockchain technology to predict an undergraduate student' future specialty in which he will more likely to succeed, by incorporating a machine learning technique known for its ability to predict and analyze data. Support vector Machine (SVM) [6] is one of these techniques, which used in this study to predict the specialized training that matches students' competencies and achievements. The SVM model discovers the relationship between the common core subjects and the specialized achievement. While, blockchain technology provides a decentralized database containing all of the data used to feed the SVM machine learning model. Furthermore, storing data on blockchain technology allows stakeholders to directly discover and obtain student data without the need for an intermediary authority. The solution proposed in our work can assist universities in reducing administrative and bureaucratic costs and efforts.

The rest of this paper is organized as follows: The second Sect. (2) is devoted to present the blockchain technology evolution and its different characteristics. Section (3) summarizes the current state of blockchain technology applications in the education field. Section (4) is reserved for the presentation of the proposed method based on blockchain technology, for the automatic classification and prediction of the future specialty of university students. Section (5) discusses the results we conducted on our system. Finally, we end this work with a conclusion.

2 Blockchain Evolution and Its Features

2.1 Evolution of Blockchain Technology

To date, blockchain technology has undergone four major evolutions (See Fig. 1) including:

Blockchain 1.0: Born from the idea of Distributed Ledger Technology (DLT) [7,8], which is a consensus-based database shared by many participants and solves the issue of double spending [9]. In 2008, crypto-currency (Bitcoin) was its most significant use [3]. It has established its stability, dependability, efficacy, simplicity, independence, and, most importantly, security to oversee transactions and transmit authority between various users directly [9]. Compared to typical payment methods, bitcoin provides a number of benefits, including reduced transaction fees and a largely anonymous payment system. Additionally, it prevents fraud by facilitating safe, open, and traceable transactions. The inability of blockchain 1.0 to allow smart contracts and the fact that it can only be used for crypto-currency applications are two of its main drawbacks. Additionally, it employs a Proof-of-Work (PoW) consensus technique [3] that necessitates intricate mathematical calculations. Additionally, research demonstrates that blockchain 1.0 has a very low throughput (maximum 7TpS).

Blockchain 2.0: *Buterin* [10] expanded the idea of blockchain beyond crypto-currencies as a result of the waste and poor scalability of blockchain 1.0. Due to this, "Ethereum," a second-generation blockchain built around the idea of smart contracts and proof-of-work compatibility mechanisms, has emerged. Stand-alone programs known as smart contracts [11] run on the basis of previously agreed-upon terms between two parties. Smart contracts considerably lower the costs of enforcement and verification, stops fraud, and allow for transparent contract identification.

Blockchain 3.0: The main shortcomings of the previous two generations are their inability to scale and excessive cost. As a result, the third generation, known as blockchain

3.0, was created with the goal of making crypto-currencies usable everywhere. Decentralized applications (dApps), which are digital programs that operate on a network of blockchain computers rather than a single machine, are the main component of this generation, along with smart contracts. Hash functions can therefore be used by this generation to enhance cross-chain transactions [12] (each node contains only part and not all of the data). By distributing the burden in this way, the system becomes effective and impervious to hacking attempts. Additionally, Proof of Stake (PoS) and Proof of Authority (PoA) consensus procedures are used in Block-chain 3.0 [13]. For earlier generations, Blockchain 3.0 intends to increase scalability, interoperability, privacy, and sustainability. It employs built-in procedures rather than relying on miners to validate and authenticate transactions. As a result, it is extremely quick because, in contrast to earlier generations, it supports thousands of transactions per second [9]. Due to the lack of a single central authority in Blockchain 3.0, there is no single point of failure, no dApp on a specific IP address (hackers/attackers cannot tamper with data), security is enhanced, and transaction speeds are extremely high.

Blockchain 4.0: Is a new generation in development that intends to make blockchain technology available as a platform for companies to create and run apps, turning blockchain into a fully integrated technology. Other technology, like artificial intelligence, can be ingrained in this generation. In order to meet business and industrial expectations, it enables the seamless integration of several platforms to operate under one roof consistently. Additionally, it has the potential to enable up to 1 million TPS of transactional speed, which is currently not attainable with earlier generations.

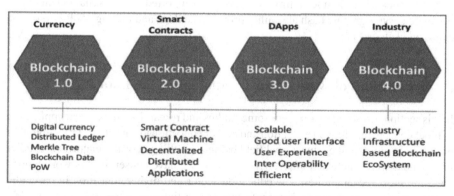

Fig.1. Blockchain evolution: From 1.0 to 4.0 [14].

2.2 Features of Blockchain Technology

Blockchain has various unique characteristics, which are not found in most other contemporary technologies, we will list the best known in the literature and the most important of them:

Immutability: The most distinctive aspect of blockchain is its improvement of tamper-proofing and provision of immutable ledgers on the P2P network. He won't be able to alter the data once it has been saved and placed on the blockchain because all other ledgers have an updated copy [15,16]. The cryptographic hash function of the blockchain is what gives it its immutability because it makes sure that even a small modification in the data will be immediately noticed [5].

Decentralization: Due to the Peer-to-Peer (P2P) design of blockchain technology, where the verification, storage, and transfer processes are based on a distributed structure, the trust between users is based on mathematical methods rather than by centralized organizations, and distributed ledger data is publicly shared across network nodes [17]. So, neither a centralized component nor a single point of failure exist [5].

According to *Vitalik Buterin*, the inventor of Ethereum, decentralization offers fault tolerance, resistance to attacks, and sedition-resistant behavior, and the blockchain is de-centralized on two of the three potential axes of software decentralization [10]:

- Politically, nobody is in charge of it.
- Logically, the system behaves as a single computer and has a generally recognized state.
- Architecturally, there is no single point of failure for the infrastructure.

Traceability: On the blockchain, transactions are listed chronologically and they are arranged in blocks that are linked together by a cryptographic fingerprint. Consequently, the capacity to look up the details of each block qualifies as the traceability of a transaction. The network-wide traceability of an event is supported by blockchain technology. A simplex cryptographic hash algorithm protects the data, and mining pools control the entire blockchain [17].

3 Applications of Blockchain Technology in the Education Field

In this section, we will try to explore some studies and research on the current application of blockchain technology in education and higher education in particular. The integration of this technology in the educational field being something new, the majority of studies in this field were published after 2016. For this, we have chosen in this work among them the most recent studies. The education sector and higher education in particular has undergone several reforms in recent years. According to the study presented by [18], the adoption of e-learning in higher education has grown steadily over the past 13 years. However, this field still faces many problems related to credibility, authenticity, security, confidentiality and transparency in sharing and retrieving any type of their data [19]. Blockchain technology has enabled educational institutions to create an infrastructure that allows them to document, store and manage their data as well as provide students with a history of their achievements that can be independently tracked and audited [20] safely and confidently. According to the study [21], the report of the joint research center (JRC) of the European commission suggests that the granting of certificates and diplomas is an important application for the field of education [22], which involves the monitoring and control of learning data to prove and validate these certificates [23]. Several researches and studies have based on the application of blockchain technology for the management (Share, transfer and verify) of certificates and diplomas [24,25,26]. Other research has focused on the student assessment and accreditation process [21,27]. However, most research has studied the protection and security of data and academic records [21,28,29].

4 Proposed Method

Higher education in Algeria still uses the principle of student classification in application of Order No. 714 of November 3, 2011 [1], in particular article 3 which stipulates on the calculation of the classification average: "the average of classification is the average of the averages of the semesters of studies concerned affected by corrective coefficients taking into account the cumulative delays, admissions with debts and admissions after the remedial session", and which is practically carried out according to the following formula:

$$MC = MSE\left(1 - a\left(r + d/2 + s/4\right)\right) \tag{1}$$

$$\text{With} \quad MSE = \sum (MSi)/n \tag{2}$$

CM: ranking average.

MSE: average of the averages of the semesters concerned (where MSi is the average of semester i).

a: abatement rate estimated at 0.04.

r: number of repetitions per year,

d: number of admissions with debt per year,

s: number of admissions after the remedial session per semester,

n: number of semesters concerned (between 1 and 6 for the License cycle and between 1 and 4 for the Master cycle).

This method does not take into account the educational data of previous students to classify and orient students [1]. Therefore, it is certainly not educational either that the average ranking is the only criterion to express the success or the failure of a student in a particular study. Moreover, this method does not reflect the fact that a particular student has actually acquired these skills and achieved these results. Because, as we mentioned earlier, centralized education systems are more vulnerable to fraud and information hacking. From this, we proposed in figure bellow (See Fig. 2) a conceptual model of a system based on blockchain technology by exploiting the power of machine learning algorithms [30] to train machine learning classifier model, in order to accurately predict the future specialty which corresponds to the students' qualifications.

It is known that the main factors which influence the success or the failure of a student in his studies of a particular discipline are: On the one hand, the degree of student' inclination, the study (training) and the subjects that compose this study. On the other hand, prerequisites in terms of subjects acquired during and before the specialty course. The figure above represents the basic principle of our system based on blockchain

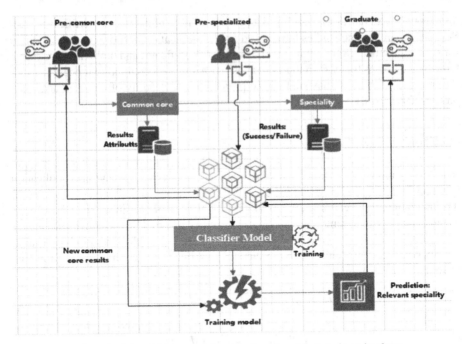

Fig.2. Principle of the proposed system based on blockchain technology.

technology, in which the student' inclination, as well as its prerequisites are expressed in a scattered way in the set of evaluation' results along its course before the specialty. Moreover, the pedagogical results of students from previous promotions are used to be able to select a relevant specialty based on the results of the student to be guided, and the results obtained by students in previous promotions. Blockchain technology has a great potential for students and universities, in designing and implementing learning activities, as well as monitoring and tracking the entire learning process [31]. Therefore, in this work the educational evaluations' results of students as well as the results of previous promotions will be stored on the decentralized database using blockchain technology to solve the problems of falsifications and fraud encountered in the educational systems still based on centralized databases.

To better understand the basic principle of the proposed system, we will give in the following subsection a scenario's example of the predicting process of a future specialty for a second-year law student as well as the different interactions with our blockchain-based system. The example of a specialty forecast appropriate to the prerequisites and the degree of inclination of the user begins with: (1) University must first register on the system by creating a wallet, in order to contact the other system members. (2) Then the university will enter all the data of their students into the system which will then register the student in the blockchain system by generating an identifier *ID* for each registered student, in our case we used the student registration number as identifier. (3) The data recorded on the blockchain is data concerning all the final results of the end-of-cycle deliberation in the various specialties, students from previous promotions, the general average of the specialty, and the mention (admitted/adjourned). (4) The student can view, verify, download or share their data recorded on the blockchain, by logging into the blockchain system using their ID. (5)Then train the supervised learning algorithm SVM [6] (Support Vector Machine) on the data and the different educational results stored on the blockchain to create a data model that can then be applied in the classification and the forecast of the future specialty of each new student of our system. The SVM classifier model was chosen because it can handle high-dimensional data and considerably improves accuracy while requiring less computational power [32]. Additionally, the SVM model employs a linear model with a class of nonlinear boundaries based on support vectors to estimate the decision function with the least amount of error [33]. (6) When a student completes his pre-specialty course, he simply logs into the proposed blockchain system using his registration number as an identifier, in order to obtain all the information on his relevant specialty proposed by the system. (7) The system will use the data model created by the SVM algorithm on the history of data and previous results recorded on the blockchain to analyze the student' learning needs, in order to predict the relevant specialty. (8) Once completed, the system extracts information about the proposed specialty and registers the result (the relevant specialty) on the blockchain, so that the students can consult or download it. (9) Instead of sending

---

Algorithm: Predicting the relevant specialty

---

1. **IN**: data: The final results of the end-of-cycle deliberation in the various specialties, students from previous promotions, the general average of the specialty, and the mention (admitted/adjourned).
2. **OUT**: Decision: Classify the student to the suitable specialty (PS, DS)
   // Registration step
3. **For** i=1 to n do   // *for each student*
4.   ID[i]= S[i]   // *Create an identifier for each student*
5. W_{BC}=hash(ID[i])   // *create a wallet for each student on the blockchain*
6. D_{BC}=hash(data[i])   // *store all data on the blockchain*
7. **End for**
   // *prediction step*
8. D=D_{BC} // *retrieve data from blockchain*
9. X= D asArray() // *create an array of student results*
10. Y= [DS, PS] // *create an array of two available specialties*
11. X_test = X*0.25 // *select 25% of dataset D for the test data*
12. Y_test= Y
13. SVM_model=SVM(X,Y)   // *training the SVM classifier model*
14. Result=SVM_model.score(X_test, Y_test)   // *testing the SVM classifier model*
15. **For** i:1 to n do
16.   Decision=SVM_model(X[i])   // *prediction the relevant specialty for the student i*
17.   Return Decision // *return the relevant specialty*
18. D_{BC}=hash(Decision)   // *Store the prediction result on the blockchain*

---

19. **End for**

Fig.3. An overview of the relevant specialty prediction steps in our system.

hard copies, the university issues electronic documents to students where each student also receives an electronic file of their learning needs and their relevant specialty (Future Specialty). (10) When the student completes his specialized studies, the university issues an electronic diploma instead of sending a hard copy to the student, who can then share it with companies for recruitment purposes, or with other universities to continue their studies.

The figure below (see Fig. 3) represents a pragmatic algorithm that offers an overview of the different steps involved in determining the relevant specialty (future specialty) of students using our proposed system.

5 Discussion and Results

This section is reserved for discussing the evaluation of the proposed system by testing its performance in predicting and classifying students at the end of the common core of the course in the different specialties concerned.

For the implementation of our system, we opted to use Python programming language, because of its power and its ease of learning. In addition, python has high-level data structures and allows for a simple and efficient approach, as well as being an ideal language for scripting and rapid application development in many areas and on most platforms.

For the learning phase, we used a dataset with 17 attributes containing 305 s-year law students at the 20th August 1955 University of Skikda, Algeria. The figure below represents the distribution of students over the 2 specialties available for the law course (DP to represent Public Law, and DS to present Private Law specialties). For the test phase, we used a dataset composed of 77 students from the same sector with 17 attributes as well. Regarding the evaluation step of our system, we built and trained the SVM model on the training dataset. Then we used the trained model to generate predictions on the test dataset. The description of the datasets used in this work is as follows:

- ID: represents the registration number of the student
- Full name
- Subject averages: were we have (CL1, CmL, STC, PbIL, FaL, M1, L1, CL2, CAPL, CPL, HR, WL, M2, L2) to represente (Civil Law1, Commercial Law, Scientific theory of crime, Public international Law, Family Law, Methodology1, Language1, Civil Law2, Civil and Administrative Procedure Law, Human rights, Work Law, Methodology2, Languge2) respectively.
- Decision (D): DP (Public Law) and DS (Private Law) represent the two specialties of the third year of the law course.

The figure below (see Fig. 4) presents an example on the training dataset used in this work.

Subsequently we created a table for the attributes which represent the 14 subjects taught and the output variable which represents the class or the specialty. In the next step, we used the trained SVM model to generate predictions on the test dataset, as follow in figure bellow (see Fig. 5).

ID	Full-name	CL1	CmL	STC	PbIL	FaL	M1	L1	CL2	CAPL	CPL	HR	WL	M2	L2	Decision
0	8	H	15.2	12.1	15.4	17.0	14.5	10.4	18.5	12.4	13.2	14.8	16.0	15.0	14.5	18.0	DS
1	9	J	15.6	17.2	14.6	19.0	16.5	13.2	11.5	17.2	13.8	14.4	9.0	8.0	13.3	14.0	DS
2	13	N	14.5	12.4	15.0	18.0	15.5	10.0	17.0	15.0	11.8	8.5	11.0	10.0	9.3	13.0	DP
3	14	O	15.7	14.0	13.0	19.5	11.0	13.2	18.0	12.6	15.8	15.7	11.0	15.0	13.5	13.0	DS
4	15	P	14.8	14.1	15.2	13.0	11.0	13.8	18.0	13.8	13.2	14.8	13.0	13.0	14.0	16.0	DP
5	16	Q	14.4	12.0	11.7	11.0	10.0	15.8	19.0	15.6	13.8	16.2	15.0	10.0	14.8	16.0	DP
6	19	T	12.8	13.4	14.5	18.0	15.0	12.8	13.0	14.0	16.1	16.1	17.0	4.0	12.7	14.0	DP
7	20	U	12.4	14.0	12.0	12.0	7.0	12.8	19.5	16.0	14.9	16.3	18.0	11.0	14.4	14.0	DS
8	23	Y	10.6	12.8	12.8	19.5	10.0	12.2	18.0	12.4	14.2	14.6	16.0	17.0	10.2	16.0	DP
9	24	Z	13.3	14.4	14.2	19.0	15.0	12.6	18.0	10.0	14.4	14.8	13.0	2.0	10.0	12.0	DP
10	25	AB	11.4	9.6	12.8	19.0	10.0	12.8	7.0	14.8	14.2	14.9	11.0	10.0	14.5	19.0	DS
11	26	AC	11.2	11.8	14.0	17.0	15.0	12.4	17.5	12.1	14.0	14.1	10.0	8.0	12.2	14.0	DP
12	27	AD	11.9	16.4	15.4	15.5	14.5	13.4	16.5	12.2	16.2	15.4	12.0	11.0	12.5	19.5	DP

Fig.4. Training data

	ID	Full name	Decision
0	139	EK	DP
1	140	EL	DP
2	141	EM	DP
3	142	EN	DP
4	143	EO	DS
5	144	EP	DP
6	146	EQ	DS
7	147	ER	DS
8	148	ES	DS
9	149	ET	DP
10	150	EU	DP
11	151	EV	DS
12	152	EW	DP

Fig.5. Prediction data generated by SVM classifier model

Once the prediction was generated, we evaluated the performance of the model in the student rankings; we calculated the test accuracy which gave us a percentage equal to 92%.

6 Conclusion

We presented along this study a system based on blockchain technology for the automatic classification of students in the different post-core specialties based on the history of educational results of previous promotions. We trained the SVM (Support Vector

Machine) model on the prediction of the relevant specialty of the third year, on a dataset composed of 424 students in the second year common core of the law sector at the University of Skikda, Algeria. We have also used blockchain technology to store all data and educational results to ensure security, transparency, immutability and traceability. Our solution can simplify processes for automatically classifying students into appropriate specialties. Consequently, this study is essential for its determination of the course of the students, finally to limit the academic failure and to allow the students to move towards the specialties of which they have the inclination and the prerequisites.

References

1. Arrete n_714 du 03 novembre 2011. http://fs.univ-skikda.dz/documents/fac/textes/arrete 714fr.pdf (2011)
2. Philip Chen, C.L., Zhang, C.Y.: Data-intensive applications, challenges, techniques and technologies: A survey on big data. Inf. Sci. **275**, 314–347 (2014)
3. Nakamoto, S.: Bitcoin: a peer-to-peer electronic cash system (2008)
4. Parino, F., Beiró, M.G., Gauvin, L.: Analysis of the bitcoin blockchain: Socio-economic factors behind the adoption. EPJ Data Sci. **7**(1), 1–23 (2018)
5. Tseng, L., Yao, X., Otoum, S., Aloqaily, M., Jararweh, Y.: Blockchain-based database in an IoT environment: challenges, opportunities, and analysis. Clust. Comput. **23**(3), 2151–2165 (2020). https://doi.org/10.1007/s10586-020-03138-7
6. Joachims, T.: Text categorization with support vector machines: learning with many relevant features. In: Nédellec, C., Rouveirol, C. (eds.) Machine Learning: ECML-98. ECML 1998. LNCS, vol. 1398, pp. 137–142. Springer, Berlin, Heidelberg (1998). https://doi.org/10.1007/BFb0026683
7. Mills, D.C., et al.: Distributed ledger technology in payments. Clearing and settlement FEDS Working Paper (2016)
8. Olnes, S., Ubacht, J., Janssen, M.: Blockchain in government: benefits and implications of distributed ledger technology for information sharing. Gov. Inf. Q. **34**(3), 355–364 (2017)
9. Mukherjee, P., Pradhan, C.: Blockchain 1.0 to Blockchain 4.0—The evolutionary transformation of blockchain technology. In: Panda, S.K., Jena, A.K., Swain, S.K., Satapathy, S.C. (eds.) Blockchain Technology: Applications and Challenges. ISRL, vol. 203, pp. 29–49. Springer, Cham (2021). https://doi.org/10.1007/978-3-030-69395-4_3
10. Buterin, V.: A next-generation smart contract and decentralized application platform. Ethereum White Pap. **3**, 1–2 (2014)
11. Macrinici, D., Cartofeanu, C., Gao, S.: Smart contract applications within blockchain technology: a systematic mapping study. Telematics Inform. **35**(8), 2337–2354 (2018)
12. Blockchain FAQ#3 : what is sharding in the blockchain? Homepage. https://medium.com/edchain/what-is-sharding-in-blockchain-8afd9ed4cff0 (2022). Accessed 21 Jul 2022
13. De Angelis, S., Aniello, L., Baldoni, R., Lombardi, F., Margheri, A., Sassone, V.: PBFT vs proof-of-authority: applying the cap theorem to permissioned blockchain. In: Italian Conference on Cyber Security. Milan, Italy, p.11 (2018)
14. IBAX Network. https://ibaxnetwork.medium.com/blockchain-evolution-from-1-0-to-4-0-18aa9ca2dbbb (2023). Accessed 03 Feb 2023
15. Karale, A., Khanuja, H.: Implementation of blockchain technology in education system. Int. J. Recent Technol. Eng. **8**(2), 3823–3828 (2019)
16. Razzaq, A., et al.: Use of blockchain in governance: a systematic literature review. Int. J. Adv. Comput. Sci. Appl. (IJACSA) **10**(5), 685–691 (2019)

17. Chen, G., Xu, B., Lu, M., Chen, N.-S.: Exploring blockchain technology and its potential applications for education. Smart Learn. Environ. **5**(1), 1–10 (2018). https://doi.org/10.1186/s40561-017-0050-x

18. Garg, A., Sharmila, A., Kumar, P., Madhukar, M., Loyola-González, O., Kumar, M.: Blockchain-based online education content ranking. Educ. Inf. Technol. **27**, 4793–4815 (2021). https://doi.org/10.1007/s10639-021-10797-5

19. Fauziah, Z., Latifah, H., Omar, X., Khoirunissa, A., Millah, S.: Application of blockchain technology in smart contracts: a systematic Litterature Review. Aptisi Trans. Technopreneurship (ATT) **2**(1), 87–97 (2020)

20. Sean Bein, A., Graha, Y.I., Pangestu, A.P.: Pandawan website design based content management system as media E-commerce transaction. Aptisi Trans. Technopreneurship (ATT) **2**(2), 160–166 (2020)

21. Tsai, C.T., Wu, J.L., Lin, Y.T., Yeh, M.K.C.: Design and development of a blockchain-based secure scoring mechanism for online learning. Educ. Technol. Soc. **25**(3), 105–121 (2022)

22. Grech, A., Camilleri, A.: Blockchain for education. Publications office of the European Union (2017)

23. Raimundo, R., Rosario, A.: Blockchain system in the higher education. Eur. J. Invest. Health, Psychol. Educ. **11**(1), 276–293 (2021)

24. Kanan, T., Obaidat, A.T., Al-Lahham, M.: SmartCert blockchain imperative for educational certificates. In: 2019 IEEE Jordan International Joint Conference on Electrical Engineering and Information Technology (JEEIT). Amman, Jordan, pp. 629–633 (2019)

25. Liu, L., Han, M., Zhou, Y., Parizi, R.M., Korayem, M.: Blockchain-Based certification for education, employment, and skill with incentive mechanism. In: Choo, K.K., Dehghantanha, A., Parizi, R. (eds.) Blockchain Cybersecurity, Trust and Privacy. Advances in Information Security, vol.79. Springer, Cham (2020). https://doi.org/10.1007/978-3-030-38181-3_14

26. Alam, A.: Platform utilising blockchain technology for eLearning and online education for open sharing of academic proficiency and progress records. In: Asokan, R., Ruiz, D.P., Baig, Z.A., Piramuthu, S. (eds.) Smart Data Intelligence. Algorithms for Intelligent Systems. Springer, Singapore (2022). https://doi.org/10.1007/978-981-19-3311-0_26

27. Bjelobaba, G., Paunovic, M., Savic, A., Stefanovic, H., Doganjic, J., Bogavac, Z.M.: Blockchain technologies and digitalization in function of student work evaluation. Open Access J. (MDPI). Sustainability **14**(9), 5333 (2022)

28. Sunarya, P.A., Rahardja, U., Sunarya, L., Hardini, M.: The role of blockchain as a security support for student profiles in technology education systems. Jurnal Nasional Informatika dan Teknologi Jaringan **4**(2), 5 (2020)

29. Rizky, A., Kurniawan, S., Gumelar, R.D., Andriyan, V., Prakoso, M.B.: Use of blockchain technology in implementing information system security on education. J. Biol. Educ. Sci. Technol. **4**(1), 62–70 (2021)

30. The royal society, London. Machine learning: the power and promise of computers that learn by example. Homepage. https://royalsociety.org/machine-learning/. ISBN 978–1–78252–259–1 (2017)

31. Chen, G., Xu, B., Lu, M., Chen, N.S.: Exploring blockchain technology and its potential application for education. Smart Learn. Environ. **5**(1), 1—10 (2018)

32. Garc´ıa-M´endez, S., Fern´andez-Gavilanes, M., Juncal-Mart´ınez, J., Gonz´alez-Casta˜no, F. J., Seara, O. B.: Identifying banking transaction descriptions via support vector machine short-text classification based on a specialized labelled corpus. IEEE Access. **8**, 61642–61655 (2020)

33. Pambudi, N. B., Hidayah, I., Fauziati, S.: Improving money laundering detection using optimized support vector machine. In: 2019 International Seminar on Research of Information Technology and Intelligent Systems (ISRITI), pp. 273–278. IEEE (2019)

MyGov Review Dataset Construction Based on User Suggestions on MyGov Portal

Sushma Yadav and Rudresh Dwivedi[✉]

Department of Computer Science and Engineering, Netaji Subhas University of Technology (NSUT), Delhi, India
{sushma.yadav.phd21,rudresh.dwivedi}@nsut.ac.in

Abstract. MyGov (*www.mygov.in*) is a well known citizen engagement platform (CEP) launched in July 2014 by Government of India (GoI) where millions of people share their innovative ideas about different government services and schemes [1]. The aim of this study is to extract user's comments, opinion and reviews from MyGov website. As a result, we have constructed two sets of dataset that is labelled and unlabelled user reviews. The labelled one contains 2000 user comments where class label 1 represents useful information for Indian government, 0 represents neutral comments, and -1 category represents comments that contain indignity. The other dataset i.e. unlabelled dataset contains 14,912 user comments in six languages. We have utilized python packages to extract the data that is capable of facilitating the process of fetching information from MyGov, allowing for easy public knowledge of the comments that exist in specific period and serving as a source of inspiration for subsequent development. MyGov has given citizens access to a fresh facet of democracy. By empowering citizens to come up with solutions and join an integrative and participatory governance framework by sharing their perspectives on choices, policies, and programmes made by the government, it promotes the crowdsourcing of ideas from communities. Further, we present classification summary that includes the results of typical machine learning (ML) and deep learning models in single table. In turn, the acquired dataset will be a boon to government to set up policies and benchmark to research community for experimental evaluations.

Keywords: MyGov · Citizen Engagement · NLP · User reviews

1 Introduction

1.1 Background

Social media has developed as a tool for personal communication as well as an outlet for people to voice their thoughts about products, services, politics, and current events. Due to its extensive popularity, a huge number of user evaluations or opinions are created and exchanged every day. With over 4.2 billion active social media users connecting to the Internet in January 2021 [2], the amount of data and metadata are astounding. Since technology progress has accelerated recently, it is now impossible to imagine life without the internet. The enormous amount of data is generated by a variety of

A. Bennour et al. (Eds.): ISPR 2023, CCIS 1940, pp. 228–240, 2024.
https://doi.org/10.1007/978-3-031-46335-8_18

online sources, with social media users setting the bar for data production. Social media has altered the platforms on which people discuss their thoughts and perspectives, as well as their opinions on goods and services. In recent year, blogs, online forums, and Social Networking Sites (SNS) have produced vast amounts of data that may be befitting for business intelligence [3]. Individuals are social creatures by nature, and SNS platforms provide a variety of ways for people to stay in touch with one another around the globe. Investigations of user behaviour, attitudes, or usage across all SNS stages that exhibit user contributions or user actions are regarded to be a part of social media research. There are numerous methods exist for fine tuning user opinions gathered via social media. Unstructured data analysis is a difficult and getting valuable information out of the large unstructured data is a challenging issue. This research aims to extract user comments from *www.mygov.in* that may be in different data types and formats. In addition to typical data types and files from conventional statistical software packages, python can read a broad variety of advanced formats. Governments all around the world are now deploying Information and Communications Technology (ICT) to involve citizens in their operations. It has been observed that an expansion of online platforms for providing services, open data portals and complaint resolution, etc. Digital Citizen Engagement (DCE) platforms, which leverage new media/digital ICTs to build or improve communication channels to involves citizens. MyGov is a unique one-stop and dedicated indigenous social media network of the GoI and a DCE that also exists in India. However, How it operates in an under-developing countries, may limit its impact [4].

Figure 1 is a screenshot of MyGov home page. The uppermost part of the Fig. 1 depicts that MyGov inviting suggestions on India's priorites for G20 presidency 2023 and the themes for suggestions are agriculture, anti-corruption, culture, development, digital economy, disaster risk reduction, education, employment, energy transition, environment, and climate sustainability, health, tourism, trade, investments, and finance work-stream. The second part depicts that the Ministry of Electronics and Information Technology (MeitY) is launching a campaign called **Stay Safe Online** during India's G20 presidency to raise awareness among citizens includes a campaign people with special needs, about how to stay safe in an online environment given the widespread use of social media platforms and the quick uptake of digital payments. This campaign also focuses on educating users of all ages about online risk and safety measures and promoting cyber hygiene in order to reinforce people's cyber safety as India is approaching to become a trillion-dollar digital economy. Participants in the Be Safe Online campaign are invited to be people of all ages, from both urban and rural areas, with a focus on children, students, women, teachers, professors, older citizens. The lower part of the snapshot shows that the participants may take part in various activities such as do/task, discuss, poll/survey, blog, talk, quiz, campaign, podcast.

This platform aims to outline the issues from a political, social, and technical standpoint. This has provided research community to propose some solutions that might assist the government in addressing these issues. Several SNS such as Facebook, Instagram, and Twitter, are being used by thousands of individuals every seconds to exchange information such as images, locations, reviews of their surroundings, and activities. Also, this offers an opportunity to service providers. MyGov may also serve as a meet-

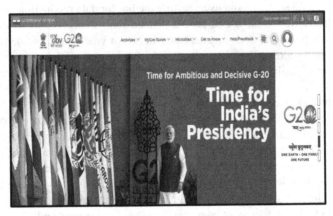

Fig. 1. An example of MyGov home page which states India's presidency G20 summit, awareness for cyber security and different categories of collecting user reviews.

ing place for important parties which includes the concerned ministry (who must evaluate public opinion or crowdsource ideas for the creation and execution of new policies) and the general public (who will be affected by the policies). MyGov initiates this communication in order to contribute meaningful rather than merely exchanging thoughts. This online political portal has 80 Groups and each represents a specific government department or sector. All these groups are organized around one of three participatory possibilities: Do, Discuss, or Disseminate [1,4,5]. These three possibilities are discussed briefly in the following paragraphs.

Do: Give Your Time to Fostering National Development. According to the interest, residents can complete the online and on-Ground tasks. There are several tasks active under each of the 80 groups under Do sections on the MyGov portal. Figure 2 is a screenshot of the **Do** section under MyGov portal.

Fig. 2. Logo design contest for stay safe online campaign.

Figure 2 shows that the Ministry of Electronics and Information Technology (MeitY) had organized logo design competition for **Stay Safe Online** in order to create awareness among citizens to stay safe online on the widespread use of social media platform and an electronic payments. After the competition's announcement, a winner entry from among the 522 received entries was announced for each group. The winning entry was duly acknowledged and is currently featured in all of Stay Safe Online communications.

Discuss: Describe Yourself. Figure 3 shows an example of discuss section under MyGov portal. This section enables citizens to share their valuable suggestions and thoughts on topical debates towards different government policy initiatives. People may participate in **group-centric** online debates on several **national themes** through public consultations and open forums by commenting on the subject and responding to other citizen's remarks in order to generate productive dialogue. In Fig. 3, the Department of Economic Affairs, Ministry of Finance invite suggestions from Indian people every year, to make the budget-making process participative, inclusive, and to promote **Jan Bhagidari's spirit**.

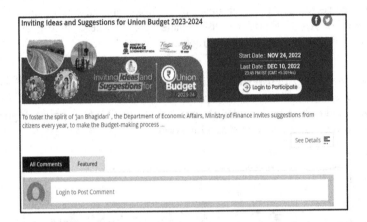

Fig. 3. Discuss section under MyGov portal.

MyGov Quiz, Polls and Surveys: Tell Us What You Think. In addition, there are online polls and surveys (referred as Janmat) that people may take part in to have their opinions without having to invest their time and energy in debates. MyGov Polls and Surveys offer residents the chance to voice their opinions on specific policy matters via online polls, providing the government with useful information regarding the success and acceptance of its policy initiatives.

1.2   Motivation and Contribution

Our proposal for dataset construction is intended for automated user-opinion acquisition, relevant reviews extraction, their analysis, classification and recommendation. The government can know that why and which policy/scheme/product must be modified or which one is actually useful or working accurately. As a result, they can improvise their scheme or policy. We collected a dataset from MyGov government website. Hence, we first constructed two sets of the data one is labelled and the other one is unlabelled dataset. Table 1 shows an overview of the dataset statistics. Distribution of categories (1, 0, -1) in labelled dataset is shown in Fig. 4 before and after applying upsampling techniques. Purple, pink, and green colour in Fig. 4 represents -1, 0, 1 categories respectively.

The outline of this paper is structured as follows: Sect. 1 describes the need of MyGov public opinions, followed by motivations and contributions. Detailed literature survey on E-Governance, machine learning, deep learning techniques, and differ-

Table 1. Overview of the dataset statistics

Dataset Specification			
Labelled Dataset		Unlabelled Dataset	
Total number of user reviews	2000	Total number of user reviews	14,912
Total number of categories	3	Total number of languages	6
Total number of languages	2		

ent data extraction methods are presented in Sect. 2. Section 3 describes experimentation methodology for data extraction from MyGov portal. Implementation and experimental results are demonstrated in Sect. 4. Further experimental evaluation on different machine learning and deep learning classifiers are demonstrated in Sect. 5. Section 6 concludes the paper.

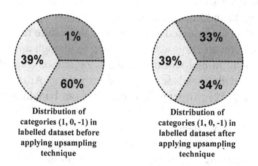

Distribution of categories (1, 0, -1) in labelled dataset before applying upsampling technique

Distribution of categories (1, 0, -1) in labelled dataset after applying upsampling technique

Fig. 4. Distribution of categories (1, 0, -1) in labelled dataset. (Color figure online)

2   Related Work

For safe and reliable information management, recent studies have concentrated on incorporating machine learning and deep learning techniques for different applications [1,6,7]. Najid et al. [8] extracted tweets related to Ibn Khaldun's thoughts. The tweet extraction has been carried out using an algorithm written in R programming language. 45 keywords based on Ibn Khaldun's thoughts were constructed as the hashtag to retrieve data from Twitter. As a result, 1075 public tweets were collected through the search API. Muttakin et al. [6] have developed an automated system (BRevML) to detect the sentiment polarity from customer reviews, the author collected the reviews from popular ecommerce websites manually. Further, the dataset is analyzed with different machine learning classifiers such as naive Bayes (NB), logistic regression (LR), support vector machine (SVM), random forest (RF), extreme gradient boosting (EGB), and deep learning classifiers such as artificial neural network (ANN), long-short term memory (LSTM), and gated recurrent units (GRU). The experimental results have shown that

SVM is more effective among all classifiers with F1-score of 89% to classify the polarity of customer Bangla reviews. Verma et al. [1] extracted the sentences from MyGov portal utilizing Stanford's sentence extractor, followed by tokenization and POS tagging. In turn, the tagged comments generate suggestions and polarity of comments. The suggestion extraction algorithm based on the observation that sentence containing suggestions are basically in the future tense. The approach achieved 86% and 88% accuracy for opinion mining and suggestion extraction, respectively. Malhotra et al. [4] addressed the challenges faced over DCE (mygov.in) platform in developing countries. In addition, the authors have analyzed three modes of participation (Do, Discuss, Disseminate) over different ministries. Ghosh [9] in her research, presented the exploratory research which highlights various opportunities for the populace as well as various difficulties the Digital India programme has encountered. Munna et al. [10] have designed a deep learning model for predicting online product quality (good or bad) and classify them into four categories (complain, recommended, wrong delivery, appreciation) based on the product reviews of the customers. They have acquired a product reviews from various e-commerce sites such as Daraz, BDshop, and Evally. Dataset contains Bangla, Phonetic Bangla and English text.

Materzynska et al. [11] presented a new large scale gesture recognition dataset. The motive of this dataset construction was to provide an enough amount of training data that could work robustly in real-world scenarios, the author created a data collection platform that interacts with crowdsourcing services such as Amazon Mechanical Turk (AMT). Kowsher et al. [12] extracted information from bangla names. Using a technique, i.e. n-gram, ML classifier such as LR, SVM, NB, k-nearest neighbors (KNN), AdaBoost, gradient boosting, decision tree, RF, and impact learning and deep learning classifiers such as ANN, LSTM, and convolutional neural network (CNN) were used for experimental evaluations. Amongst all, impact learning, AdaBoost, and gradient boosting yield the optimal results for different information extraction.

3   Methodology

The process of collecting and analyzing data on certain attributes is regarded as data extraction. The accurate data collection further ensures the integrity and validity of reviews acquired. This section identifies probable sources for MyGov comments retrieval. Usually, web scraping deals with publicly available data as the raw data may not be interpretable due to different formats. In this work, we describe that how we may extract data from any website with the help of selenium python package. As MyGov is a dynamic website that includes a **View More** button instead of pagination at the bottom of the page. The view more buttons are dynamic components that display additional data just after the existing data. Each press adds more data as long as there is still hidden data accessible. After all of the hidden data is displayed on the page, the view more button is eliminated, without utilizing selenium we may extract data only from the first page. Therefore, we have incorporated selenium to extract data from multiple pages [13].

In our work, we utilized chrome driver and import requisite libraries. Selenium Webdriver provides a controls over any web browser by directly communicating with

Fig. 5. Flow diagram of proposed method.

it. Next, we created fresh instance of chrome browser and as a result, our program would be able to open a URL in chrome browser (Figs. 5 and 6).

Fig. 6. Chrome controlled by automated test software.

Next, we use XPath locators to navigate through the page's HTML code. This facilitates to locate any element on a webpage in HTML/XML documents. The path of an element located on the web page is contained in XPath. There exists two categories for XPath locators (i.e. absolute XPath and relative XPath). In our work, we have used absolute XPath as it contains the entire path from the root element, the location of the element would be faster when compared with other types of XPath. The general Syntax for **XPATH** is given as follows:

$$XPATH = //tagname[@attribute =' value'] \tag{1}$$

```
Xpath=//input[@type='text']
Xpath=//label[@id='Sushma93']
Xpath=//input[@value='RESET']
Xpath=//*[@class='comment_body']
Xpath=//a[@href='http://demo.democlass.com/']
Xpath=//img[@src='//democlass.com/images/home/sushma.png']
```

Fig. 7. Multiple ways of defining XPath expressions.

Table 2. Different ways for finding an element on the web page

SNo.	XPath Locators	Find Different Elements on Web Page
1	ID	To find an element by ID of the element.
2	Classname	To find an element by Classname of the element.
3	Name	To find an element by Name of the element.
4	Link text	To find an element by text of the link.
5	XPath	XPath required for finding the dynamic element and traverse between various elements of the web page.
6	CSS Path	CSS path also locates elements having no name, class or id

Different ways for finding an element on the webpage are illustrated in Table 2. The example definitions for XPath Expression are shown in Fig. 7.

In the last step of the proposed methodology, the gathered data saved as a .CSV file.

4   Experimental Results

In labelled dataset, the -1 category contains total 13 data points, 1 category contains 1204 data points and 0 category contains 783 data points in 2 languages that is English and Hinglish. The unlabelled dataset contains 14,912 reviews in 6 languages (English, Hindi, Hinglish, Gujrati, Bangauli, Telgu). MyGov comments on a variety of subjects including **national youth policy, call for ideas-citizen interface for national highway authority of India (NHAI), inviting suggestions for proposed new cooperative policy, ideas for panchayat vision 2047, inviting Suggestions for Beach cleaning activities, and inviting suggestions for mann-ki-baat by prime minister office and/or different ministries/government bodies** are returned by the data extraction process. Two sets of dataset have been prepared; one is tagged and manually labelled into three categories (1,0,-1). The category 1 represents that comment is useful suggestion for the government organization, -1 represents that the comment containing indignity and 0 category represents that the comment is neutral (neither contains useful suggestion nor indignity). Other dataset is unlabelled and comprises 14,912 comments in multiple languages. Finally, the data consisting of user's feedback, opinion, reviews, and suggestions are exported as a data-frame.

The gathered data may contain hashtags, URLs, abbreviations, punctuation, stop words, and other symbols as shown in Fig. 8, Fig. 9, and Fig. 10. Also, few comments may have a lot of characters that are difficult to read and comprehend. Hence, further pre-processing is required to deal with missing and erroneous reviews followed by the removal of above-stated irregularities present in the reviews. Before feeding the data to any classifier, the missing, erroneous, or irrelevant elements need to be identified and corrected. The effectiveness of the proposal is that the gathered user reviews dataset on different government schemes and social issues will be a boon to government to set up new policies and benchmark to research community for experimental evaluations (Fig. 11).

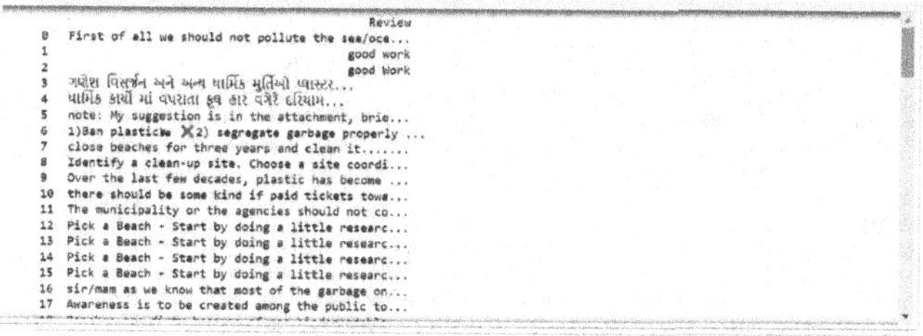

	Review
0	First of all we should not pollute the sea/oce...
1	good work
2	good Work
3	ગણેશ વિસર્જન અને અન્ય ધાર્મિક મુર્તિઓ પ્લાસ્ટર...
4	ધાર્મિક કાર્યો માં વપરાતા ફૂલ હાર વગેરે દરિયામ...
5	note: My suggestion is in the attachment, brie...
6	1)Ban plastic X2) segregate garbage properly ...
7	close beaches for three years and clean it.......
8	Identify a clean-up site. Choose a site coordi...
9	Over the last few decades, plastic has become ...
10	there should be some kind if paid tickets towa...
11	The municipality or the agencies should not co...
12	Pick a Beach - Start by doing a little researc...
13	Pick a Beach - Start by doing a little researc...
14	Pick a Beach - Start by doing a little researc...
15	Pick a Beach - Start by doing a little researc...
16	sir/mam as we know that most of the garbage on...
17	Awareness is to be created among the public to...

Fig. 8. Sample of MyGov data extraction in csv file.

5 Classification Summary

5.1 Further Experimental Evaluation on Different Machine Learning and Deep Leaning Classifiers

In this section, the experimental evaluations involving different ML classifiers have been presented for the acquired dataset. In addition, we also perform the experimentation over deep neural architectures. We apply different feature extraction techniques such as CountVectorizer, Char level TF-IDF, Word level TF-IDF, and Ngram level TF-IDF before the data is fed to classifiers. Out of all, RF using Char level TF-IDF outperforms other classifiers as it is less prone to over-fitting. The problem of over-fitting may occur in decision trees that memorize the training data as well as noisy data. This may lead to erroneous prediction. The ML models yield optimal performance than deep neural architectures in scenarios where the size of dataset is small. We have evaluated three different performance metric i.e. precision, recall, and F1-score for different classifiers as reported in Table 3.

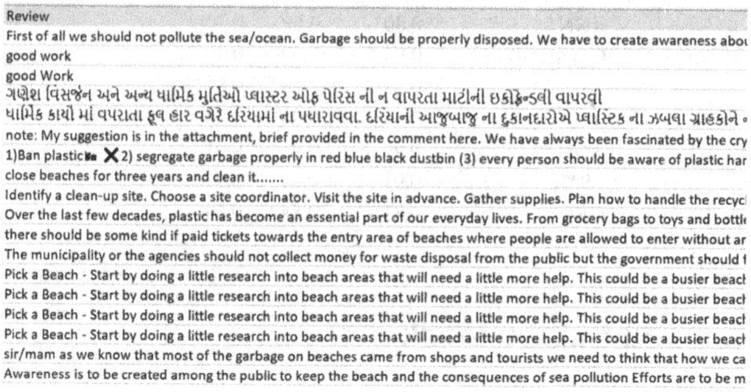

Review
First of all we should not pollute the sea/ocean. Garbage should be properly disposed. We have to create awareness abo
good work
good Work
ગણેશ વિસર્જન અને અન્ય ધાર્મિક મુર્તિઓ પ્લાસ્ટર ઓફ પેરિસ ની ન વાપરતા માટીની ઇકોફ્રેન્ડલી વાપરવી
ધાર્મિક કાર્યો માં વપરાતા ફૂલ હાર વગેરે દરિયામાં ના પધારાવવા. દરિયાની આજુબાજુ ના દુકાનદારોએ પ્લાસ્ટિક ના ઝબલા ગ્રાહકોને
note: My suggestion is in the attachment, brief provided in the comment here. We have always been fascinated by the cry
1)Ban plastic X 2) segregate garbage properly in red blue black dustbin (3) every person should be aware of plastic har
close beaches for three years and clean it.......
Identify a clean-up site. Choose a site coordinator. Visit the site in advance. Gather supplies. Plan how to handle the recyc
Over the last few decades, plastic has become an essential part of our everyday lives. From grocery bags to toys and bottl
there should be some kind if paid tickets towards the entry area of beaches where people are allowed to enter without ar
The municipality or the agencies should not collect money for waste disposal from the public but the government should
Pick a Beach - Start by doing a little research into beach areas that will need a little more help. This could be a busier beac
Pick a Beach - Start by doing a little research into beach areas that will need a little more help. This could be a busier beac
Pick a Beach - Start by doing a little research into beach areas that will need a little more help. This could be a busier beac
Pick a Beach - Start by doing a little research into beach areas that will need a little more help. This could be a busier beac
sir/mam as we know that most of the garbage on beaches came from shops and tourists we need to think that how we ca
Awareness is to be created among the public to keep the beach and the consequences of sea pollution Efforts are to be m

Fig. 9. Unlabelled multi lingual dataset.

1	Reviews	Category
2	sir,when automatic toll collection is carried out by using fastag. the customer can be contacted aft	1
3	Cont'd---To be a World class we need to have charging stations with Solar farm, Windmill, hydel in	1
4	An integrated app must be designed having integrated with a toll free helpline number can be pro	1
5	Recently I saw one bike rider lying spot dead on Silkboard to Electronic City National Highway. Eve	1
6	An integrated app must be designed having all information about the amenities available on NHAI	1
7	All High way is new generation High speed on the way daily routineing all languages Flag tagline on	0
8	Any intelligent person saw this comments you must see attached file you got government work	0
9	The facilities such as electric charging stations should be provided which will help in promoting us	0
10	NHAI citizen interface should be user friendly weather it may be a teenager or a women commute	1
11	Dear Modi ji Some more suggestions here as follows 1. Put Proper pavement marking, chevron sig	1
12	NHAI decides the fuel consumption & travel time. Shorter the distance with least gradients & turn	1
13	I am from Ahmednagar - Bhingar, location NH222. Here is undivided double lane road plus with fu	1
14	The heavy vehicles do not follow the lane rules and sometimes causes accidents. The App may int	1
15	In some areas there are Animal Crossing will be there. In such areas the App can give notifications	1
16	In some areas while traveling in pitch darkness during night time there are no Signboards to check	1
17	The local Police along NH irrespective of the state/UT can also be included in the 'app'	1
18	Most modern android phones have an accelerometer which can detect how much the phone is sh	1
19	National Safety Day 2021: 'Sadak Suraksha' theme to be observed this year. This year, the focus of	0

Fig. 10. Labelled dataset.

	Reviews	Category
308	respected pm sir namste thanks contribution ne...	0
1174	suggest students read	0
766	indian national highways road safety idea 7 km...	0
316	nhai good institute good work	0
234	love india	0

Fig. 11. Data after removing all types of irregularities.

Table 3. Precision, Recall, F1-score values for predicting informative reviews using classifiers.

Classifiers	Precision	Recall	F1-Score
NB	0.80	0.77	0.74
SVM	0.91	0.91	0.90
KNN	0.80	0.69	0.70
Gradient Boosting	0.90	0.90	0.90
LR	0.88	0.88	0.87
RF	0.94	0.94	0.94
ANN	0.92	0.91	0.91
LSTM	0.91	0.91	0.91
GRU	0.85	0.84	0.84

6   Conclusion and Future Work

In our work, the data from the MyGov discuss comments are retrieved and analyzed over different search phrases for enormous amounts of information. The approach is capable of retrieving and processing the user comments with ease. The dataset contains the reviews/comments/suggestions over different aspects such as national youth policy, call for ideas-citizen interface for national highway authority of India (NHAI), inviting suggestions for the proposed new cooperative policy, ideas for panchayat vision 2047, inviting suggestions for beach cleaning activities, and inviting suggestions for mann-ki-baat by prime minister office and/or different ministries/government bodies. The retrieved information may be turned into a data frame and a word cloud, that focuses on frequently occurring words in a sentence. We also plan to extend rigorous evaluation of the acquired dataset under different experimental setting. The proposed technique may further be used to acquire any other dataset, analyze various social media platforms, and show how to create cutting-edge methods for text representation in social media by using examples from the actual world. Creating a social media tracking and analysis method may be a emerging viewpoint for future work as opinions evolve over time. It may also be feasible to incorporate text mining and perform sentiment analysis, which categorise the collected information as positive, negative, or neutral which may influence decisions later.

References

1. Verma, S., Ramamurthy, A.: Analysis of users' comments on political portal for extraction of suggestions and opinion mining. In: Proceedings of the International Conference on Advances in Information Communication Technology & Computing, pp. 1–4 (2016)
2. Yerramilli, R., Swamy, N.K.: Framework for citizen adoption of egovenance services in developing countries. In: 2018 IEEE Conference on e-Learning, e-Management and e-Services (IC3e), pp. 155–160. IEEE (2018)
3. Nagarathna, R., Manoranjani, R.: An intelligent step to effective e-governance in India through e-learning via social networks. In: 2016 IEEE 4th International Conference on MOOCs, Innovation and Technology in Education (MITE), pp. 29–35. IEEE (2016)
4. Malhotra, C., Sharma, A., Agarwal, N., Malhotra, I.: Review of digital citizen engagement (DCE) platform: a case study of mygov of government of India. In: Proceedings of the 12th International Conference on Theory and Practice of Electronic Governance, pp. 148–155 (2019)
5. Misra, A., Misra, D.P., Mahapatra, S.S., Biswas, S.: Digital transformation model: analytic approach on participatory governance & community engagement in India. In: Proceedings of the 19th Annual International Conference on Digital Government Research: Governance in the Data Age, pp. 1–7 (2018)
6. Alam, M.M., Shome, A., Saha, S., Mridha, M.: BRevML: classifying Bangla reviews for e-commerce using machine learning. In: 2021 International Conference on Science & Contemporary Technologies (ICSCT), pp. 1–6. IEEE (2021)
7. Lamba, A., Yadav, D., Lele, A.: Citizenpulse: a text analytics framework for proactive e-governance-a case study of mygov.in. In: Proceedings of the 3rd IKDD Conference on Data Science 2016, pp. 1–2 (2016)
8. Najid, M.H.M., Zulkifli, Z., Othman, R., Rokis, R.: Extracting tweets using R in the context of ethical issues. In: 2021 Fifth International Conference on Information Retrieval and Knowledge Management (CAMP), pp. 41–45. IEEE (2021)
9. Ghosh, K.: Opportunities and challenges of digital India. IJFMR-Int. J. Multidisciplinary Res. **4**(6) (2022)
10. Munna, M.H., Rifat, M.R.I., Badrudduza, A.: Sentiment analysis and product review classification in e-commerce platform. In: 2020 23rd International Conference on Computer and Information Technology (ICCIT), pp. 1–6. IEEE (2020)
11. Materzynska, J., Berger, G., Bax, I., Memisevic, R.: The jester dataset: a large-scale video dataset of human gestures. In: Proceedings of the IEEE/CVF International Conference on Computer Vision Workshops (2019)

12. Kowsher, M., Sanjid, M.Z.I., Das, A., Ahmed, M., Sarker, M.M.H.: Machine learning and deep learning based information extraction from Bangla names. Procedia Comput. Sci. **178**, 224–233 (2020)
13. Chapagain, A.: Hands-on web scraping with Python: perform advanced scraping operations using various Python libraries and tools such as Selenium, Regex, and others. Packt Publishing Ltd (2019)

IDA: An Imbalanced Data Augmentation for Text Classification

Asma Siagh[1($\boxtimes$)], Fatima Zohra Laallam[1], Okba Kazar[2,3], Hajer Salem[4],
and Mohammed Elhacene Benglia[5]

[1] Laboratoire d'INtelligence Artificielle et des Technologies de l'Information
(LINATI), Department of Computer Science and Information Technologies,
Kasdi Merbah University Ouargla, Ouargla, Algeria
`siagh.asma@univ-ouargla.dz`
[2] Smart Computer Science Laboratory (LINFI), Computer Science Department,
University of Biskra, Biskra, Algeria
`o.kazar@univ-biskra.dz`
[3] Department of Information Systems and Security, College of Information
Technology, United Arab Emirate University, Al Ain, United Arab Emirates
`o.kazar@uaeu.ac.ae`
[4] Pôle R&D, Audensiel Technologies, Ile-de-France, Paris, France
`h.salem@audensiel.fr`
[5] Laboratoire de Genie Electrique (LAGE), Department of Computer Science and
Information Technologies, Kasdi Merbah University Ouargla, Ouargla, Algeria
`benglia.elhacen@univ-ouargla.dz`

Abstract. With the increasing amount of textual data generated online,
an automatic system for text classification is imperative. However, classification models face the challenge of limited and imbalanced data, resulting
in poor performance on minority classes. This paper presents a data augmentation technique for imbalanced text classification called Imbalanced
Data Augmentation (IDA). The proposed technique consists of three main
components: word selection, synonym substitution, and stop word insertion. We evaluate IDA's performance using an imbalanced dataset of user-
generated feedback on Algerian higher education sourced from tweets. Our
proposed technique significantly improves the detection of the minority
class by achieving the highest F1-score compared to the other evaluated
data augmentation methods. Overall, IDA is a useful tool for enhancing
the performance of text classifiers on imbalanced datasets by preventing
overfitting, improving model generalization, addressing class imbalances,
and reducing the cost of collecting and labeling data.

Keywords: Natural language processing · Text classification · Data
augmentation · Imbalanced data

1 Introduction

Text Classification (TC) is a fundamental task in natural language processing
(NLP) that aims to automatically categorize a given text into one or more pre-
defined classes. TC is a vast research area with many subfields, such as sentiment

A. Bennour et al. (Eds.): ISPR 2023, CCIS 1940, pp. 241–251, 2024.
https://doi.org/10.1007/978-3-031-46335-8_19

analysis, topic modeling, and spam detection, to name a few. In the era of the digital world, the rapid growth of textual content has made the development of automatic and effective TC systems imperative. Due to their ability to automatically learn representative features from raw input data, deep learning models have achieved state-of-the-art performance in various NLP tasks, including TC. However, the performance of these models heavily relies on the size of training data, such that larger training datasets tend to yield better performance results. On the other hand, collecting and annotating sufficient training data is a costly and time-consuming process. Furthermore, available data can be limited and imbalanced, resulting in some classes having significantly more examples than others. This can lead to a biased model that overfits to the majority classes and performs poorly on the minority classes [9].

One of the solutions to this challenge is Data Augmentation (DA). DA is a process of increasing data size by generating new variations upon already existing data. DA first emerged in the field of computer vision, where image data is augmented using cropping, flipping, rotation, and color adjusting [8]. Following its success in computer vision, DA has gained interest among researchers across various domains, including NLP. Several approaches are proposed to text data augmentation, including text transformation, back-translation, and generative models [7]. Text transformation refers to any modification made to the original text while preserving its meaning. Synonym replacement is one of the commonly used techniques, where words in a given text are substituted with their synonyms based on WordNet [3], pre-trained word embeddings, or masked language models. WordNet is a lexical database that stores words and their relationships, while pre-trained word embeddings are a type of language model that maps words to vectors in a high-dimensional space. Masked language models, on the other hand, are a type of neural network that predict a masked word based on its context. Other techniques for text transformation include adding, deleting, or repositioning words. Back-translation is the process of translating a text from its original language to another and then back to the source language using machine translation. More complex techniques use generative models to create new data, such as language models [16] and generative adversarial networks [14].

Text transformation is a simple yet effective augmentation technique for improving the performance of deep learning models. However, one of its most common methods, random synonym replacement, can sometimes have an adverse effect on the performance of the model if the replaced words could potentially remove important information relevant to the class label. To address this issue, an Imbalanced Data Augmentation (IDA) has been proposed in this study. Instead of replacing random words, IDA selects the words to be replaced based on their relevance to the minority class label using $tfidf$ scores. Subsequently, more noise is injected into the data by inserting stop words in each sentence. The rest of the paper is structured as follows. Section 2 provides an overview of related works on text data augmentation. Section 3 explains the proposed IDA in detail. Section 4 presents the experimental evaluation, including the dataset used and the results and discussion. Finally, Sect. 5 concludes the paper.

2   Related Work

In recent years, the adoption of DA has witnessed a remarkable turnout in the field of NLP. Numerous studies have been conducted to tackle the problem of limited labeled data and class imbalance in order to improve the performance of NLP models.

The majority of studies used simple text transformation techniques and augmented data by manipulating the words in the original sentences. One famous work in this regard is EDA [13], in which four text transformation operations have been proposed involving synonym replacement using Wordnet, random word insertion, random word swap, and random word deletion. However, despite these operations being simple for DA, they can cause a loss of information and change the input sequence's label [6]. Based on this limitation, the study of [5] used the same experimental setting as in EDA and augmented sentences by the random insertion of different punctuation marks into random positions in the source sentences.

Moreover, with the aim to preserve the sentence label, the authors in [17] used the attention mechanism to select relevant words for label prediction instead of manipulating words at random. For the same goal, [15] used a conditional masked language model that randomly masks words in the original sentences and predicts relevant words to the sentence label based on both its context and the label of the sentence. Another study in [12] proposed three text augmentation methods based on word substitution. The proposed methods randomly replace a subset of words in a given text with substitutes from predetermined lists. The latter are generated using WordNet and SenticNet based on cognates, antonyms, and antipodes.

Further works relied on text translation and paraphrasing methods to generate new sentences. For example, the authors in [2] utilized a transformer-based model to translate the data into Dutch and back to English. Subsequently, the paraphrasing is performed using an English-French translation model to convert data from English to French, followed by a mixture of experts that translated back the resulting French data to English. Nevertheless, these methods can present limitations such as inaccurate translations and being time-consuming with large datasets. To improve text augmentation performance, some researchers adopt a hybrid approach that combines several methods. For instance, in [10], augmented sentences are created by replacing words in the original sentences using contextual and non-contextual embeddings, modifying the random word replacement and insertion operations of the EDA technique, translating sentences into one or two intermediate languages, and then translating them back to their source language. Another work in [4] designed two augmentation methods. The first method utilizes a generative model to create a summarized sentence by combining three input sentences as augmented data. While the second method generates augmented sentences by randomly permuting specific words' characters in the original sentences.

While DA has proved its effectiveness in improving NLP models, it is still an underexplored area in comparison to computer vision. In this study, we offer a simple and efficient solution to augment limited and imbalanced text data in a way that preserves label information and increases the classification of the minority class.

3   IDA

This section depicts our Imbalanced Data Augmentation (IDA) technique, as shown in Fig. 1. IDA aims to address the problem of deep learning models' poor performance on small and imbalanced datasets. It is composed of three components. The first component is word selection, which selects the words that should be replaced in the synonym substitution component. The latter replaces the original words with their synonyms based on Glove pre-trained word embedding. In the third component, stop words are inserted randomly in each data sentence.

3.1   Word Selection

In an imbalanced classification, the minority class is more significant, making its misclassification more susceptible than the majority class. Accordingly, the purpose of IDA is to generate augmented sentences without altering the class label of the original sentences, especially those of the minority class. Replacing a sentence's original words with their synonyms is among the commonly used data augmentation methods. However, this random selection can influence the classification results if the selected words are relevant for predicting the class label. In light of this, we use the term frequency-inverse document frequency $tfidf$ to select the less relevant words to the minority class. The point of choosing uninformative words from the minority class is due to the fact that performing $tfidf$ on the whole dataset may result in selecting relevant words for the minority class since this latter constitutes a smaller portion of the entire data vocabulary.

Word selection in IDA extracts the minority class mc from the training data. For each sentence s belongs to mc, we propose to compute the relevance of w to the polarity of s following $tfidf$ formula 1 as follows:

$$tfidf(w, s, mc) = tf(w, s) * idf(w, mc) \qquad (1)$$

with w representing words, s representing sentences, and mc representing minority class sentences.

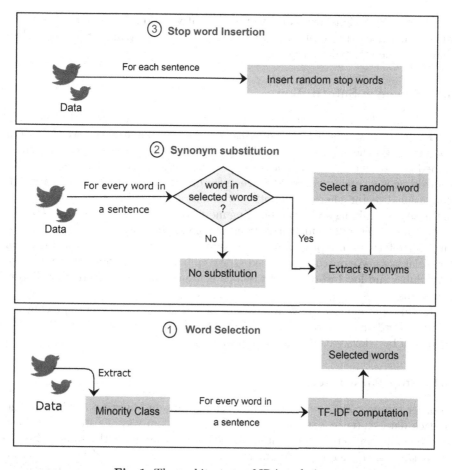

Fig. 1. The architecture of IDA technique

The term tf is a measure of how many times every word occurred in every sentence, and it is defined as follows:

$$tf(w, s) = \frac{freq(w)}{|s|} \tag{2}$$

where $freq(w)$ is the number of occurrences of the word w, and $|s|$ is the number of all words in the sentence s.

As for the term idf, it measures how common or rare is a word w across all sentences of mc, such that rare words have more value compared to common ones. The idf metric is formulated by the following equation:

$$idf(w, mc) = \log \frac{|mc|}{1 + N} \tag{3}$$

where $|mc|$ represents the total number of sentences in mc and N is the total number of sentences containing the word w.

Since $tfidf$ assigns low scores to uninformative words, we define a threshold value t that is the mean of all $tfidf$ values, and words having a $tfidf$ score less than t will be selected for replacement.

3.2 Synonym Substitution

We examine whether each word w in sentence s of the training dataset is included in the selected words list. No substitution will be made if the word w does not belong to the selected words. In the other case, we use Glove pre-trained embedding to extract the synonyms of w and perform the substitution. Word embedding is a category of word representations. It maps every word to a multi-dimensional vector wherein semantically similar words will be close to each other in the space. Training a word embedding model on small and imbalanced data for synonym extraction cannot derive valid synonyms because of the poor vocabulary. Therefore, we use pre-trained embedding to leverage the rich vocabulary of large datasets. We opted pre-trained Glove embedding due to its ability to capture global and local word context, which is important for accurately predicting synonyms.

In this component, the generation of new sentences concerns replacing the original word w by a random word out of its top fifty synonyms (N = 50) predicted by Glove with the highest probabilities.

3.3 Stop Word Insertion

In order to inject more noise into the training data and increase its size, we add to each data sentence a set of stop words in a random way. As stop words have no value and impact on the sentence's polarity, we perform the same principle of AEDA (An Easier Data Augmentation) technique proposed in [5], where a random insertion of punctuation marks is applied to the original text while preserving its information. The stop words are selected randomly from a defined stop word list and inserted in random positions into the sentence. The stop word list used is from nltk library[1] for English, French, and Arabic.

For every input sentence, we generate Ng augmented sentences, which means synonym substitution and stop word insertion are repeated $Ng/2$ each. Furthermore, to avoid adding too much noise to sentences and adversely affecting the models' learning, the number of how many words to replace and stop words to insert is controlled by the parameter n, which is calculated by $n = \alpha \times l$ with α being the ratio of substitution/or insertion and l being the length of the input sentence [13]. Table 1 illustrates an example of an augmented sentence using IDA technique.

---

[1] https://www.nltk.org/search.html?q=stopwords&check_keywords=yes& area=default.

Table 1. Application of IDA on a training sample of the dataset. The word "university" is replaced by the synonym "collage". "vos", " من ", and "in" are French, Arabic, and English stopwords.

IDA operation	Sentence
None	university residency worker stealing students food disgusting behavior
Synonym substitution	**collage** residency worker stealing students food disgusting behavior
Stop word insertion	university **vos** residency worker stealing من students food disgusting **in** behavior

4 Experimental Evaluation

In the evaluation process of IDA, we use an imbalanced dataset of tweeters' feedback on Algerian higher education. The dataset has been collected from Twitter in a recent work [11]. It has a multilingual nature and has been categorized into negative and positive classes. The former includes negative tweets constituting the minority class with 16% (2924 tweets) of the total data, whereas the latter represents the majority of the tweets with a portion of 84% (15094 tweets). For each data sentence, a series of preprocessing steps were applied, including converting all words to lowercase, expanding contractions, and removing unnecessary elements such as extra whitespace, links, numbers, stop words, punctuation, and special characters.

After the application of IDA on the dataset, we combine the original data with the new augmented ones. The resulting combination is fed as input to a deep learning model for binary sentiment classification. We use the Attention-based Bidirectional CNN-RNN Deep Model (ABCDM) from [1], which is based on Convolutional Neural Network (CNN), attention mechanism, bidirectional Long Short Term Memory (LSTM), and Gated Recurrent Unit (GRU) along with Glove pre-trained word embedding. The quality of the DA techniques is tested and compared based on the performance of ABCDM.

4.1 Evaluation Metrics

This section outlines the evaluation metrics used to assess the performance of ABCDM.

Precision measures the model's ability to correctly predict the positive class. In other words, it calculates the number of relevant tweets that the model predicted correctly. Precision is calculated using the following equation:

$$Precision = \frac{True\ Positive}{(True\ Positive + False\ Positive)} \tag{4}$$

Recall considers the number of results the model failed to classify correctly. It measures the model's ability to correctly predict both positive and negative classes. Recall is calculated using the following equation:

$$Recall = \frac{True\ Positive}{(True\ Positive + False\ Negative)} \tag{5}$$

F1-Score is a commonly used metric that combines precision and recall to evaluate the overall performance of a model and it is defined by the following equation:

$$F1 - score = 2 * \frac{Precision * Recall}{(Precision + Recall)} \tag{6}$$

Using these evaluation metrics provides a comprehensive analysis of the model's performance, ensuring that both positive and negative classifications are accounted for.

4.2   Data Augmentation Techniques for Comparison

As a baseline, we use the original imbalanced data with no DA to evaluate the augmentation effect, and we denote it as *None*. Another technique we use for comparison is *AEDA* which creates new sentences by adding a set of punctuation marks randomly to original sentences. This method proved its efficiency by outperforming EDA in the same experimental setting. Furthermore, we use different combinations of IDA operations as independent DA techniques to enhance the experimental comparison. *SWI*, for Stop Word Insertion, creates new sentences by randomly inserting stop words. *WSR* refers to Word Selection and synonym Replacement. In this operation, uninformative words from the entire dataset are selected based on $tfidf$ measure. Then, the selected words are substituted using synonyms predicted by Glove pre-trained word embedding. *IWSR*, which stands for Imbalanced Word Selection and synonym Replacement, uses the same principle as *WSR* except for selecting uninformative words from the minority class sentences. The last technique, denoted by *IWSR+SWI*, combines the operations *IWSR* and *SWI*.

In all the DA techniques, including IDA, we use α and Ng with values 0.2 and 4, respectively. These values are determined based on the recommendations stated in [13].

4.3   Results and Discussion

In this section, we present the evaluation results of ABCDM after applying IDA and the aforementioned DA techniques to the training dataset. We train ABCDM with 80% of the dataset using the same hyperparameters as in its original paper

[1], except for the epoch number, where we use 25 instead of 15. On the other hand, the remaining 20% of the dataset is used for the evaluation process in terms of Recall, Precision, and F1-score metrics. For each experiment, we run ABCDM five separate times, and the results shown in Table 2 present the average result of these runs. It is worth noting that as the minority class is considered more important in imbalanced data scenarios, we evaluate the performance based on the F1-score results of the minority class.

As shown in Table 2, ABCDM with no DA (*None*) shows the lowest result in detecting the minority class samples with a Recall of 66,91%, and this returns to the limited examples of the minority class. On the other hand, *IDA* and *WSR* highly enhance ABCDM performance in detecting the minority class with a Recall improvement of about 8% compared to *None*. As for the model precision, ABCDM with *None* is the most accurate, such that it manages to predict 79% of minority examples correctly. As Recall and Precision are inversely proportional, F1-score measure is often adopted for evaluating imbalanced classification since it considers both Recall and Precision. The best F1-score is reached by *IDA* with 74,76% exceeding *None* and *AEDA* with a marge of 2,42% and 0,37%, respectively. We also notice that using AEDA and SWI gives the same F1-score, which may be explained by their similar principle of inserting uninformative data into sentences. Furthermore, word replacement that is based on selecting words from the minority class (*IWSR*) leads to a higher F1-score in comparison to word selection from the entire dataset (*WSR*). This proves our assumption that selecting uninformative words from the whole data could affect the relevant words in the minority class. In addition, it can be noticed that the results of *WSR* and *SWI* individually are higher than their combination's result (*IWSR+SWI*). On the other hand, IDA, which combines *SWI* and *IWSR*, increases the F1-score result compared to when applying these techniques independently. This can be explained by the fact that adding and replacing uninformative words without altering the meaning of the original sentences ensures the preservation of the input information.

The main limitation of this study is that the improvement achieved by IDA may be considered modest, which is a common limitation of text transformation techniques for data augmentation. Although IDA outperformed other augmentation techniques in our experiments, the observed improvement may not be convincing enough. To address this limitation, we plan to investigate the incorporation of IDA with other data augmentation methods to further enhance its efficiency and performance. Furthermore, using the mean of *tfidf* scores as a threshold to identify irrelevant words to the minority class may not be an optimal choice. Therefore, we plan to explore alternative threshold values in future experiments to improve results.

Table 2. Comparison results (%) of ABCDM with and without the DA techniques

DA techniques	Recall	Precision	F1-score
None	66.91	**79.05**	72.34
AEDA [5]	74.49	74.42	74.39
SWI	71.02	78.13	74.39
WSR	**75.22**	72.56	73.83
IWSR	72.40	76.1	74.21
WSR+SWI	72.65	72.97	72.76
IDA	**75.13**	74.38	**74.76**

5   Conclusion

In this study, we proposed an Imbalanced text Data Augmentation (IDA) technique for boosting the prediction of minority class samples. The experimental results on an imbalanced dataset of Twitter users' feedback on Algerian higher education showed that IDA improved the detection of minority tweets and outperformed other data augmentation techniques. IDA increases the data size at two levels: firstly, enhancing the size of the minority class by replacing uninformative words while preserving the class label, and secondly, augmenting the entire data size by inserting stop words into each data sentence.

Our future work will focus on investigating the optimal number of words to be replaced and punctuation marks to be inserted to further improve the performance of IDA. Additionally, we plan to test IDA on benchmark datasets to evaluate its effectiveness.

References

1. Basiri, M.E., Nemati, S., Abdar, M., Cambria, E., Acharya, U.R.: ABCDM: an attention-based bidirectional CNN-RNN deep model for sentiment analysis. Futur. Gener. Comput. Syst. **115**, 279–294 (2021)
2. Bayer, M., Kaufhold, M.A., Buchhold, B., Keller, M., Dallmeyer, J., Reuter, C.: Data augmentation in natural language processing: a novel text generation approach for long and short text classifiers. Int. J. Mach. Learn. Cybern., 1–16 (2022)
3. Fellbaum, C.: Wordnet. the encyclopedia of applied linguistics (2012)
4. Jo, B.C., Heo, T.S., Park, Y., Yoo, Y., Cho, W.I., Kim, K.: Dagam: data augmentation with generation and modification. arXiv preprint arXiv:2204.02633 (2022)
5. Karimi, A., Rossi, L., Prati, A.: Aeda: an easier data augmentation technique for text classification. arXiv preprint arXiv:2108.13230 (2021)
6. Kumar, V., Choudhary, A., Cho, E.: Data augmentation using pre-trained transformer models. arXiv preprint arXiv:2003.02245 (2020)
7. Li, B., Hou, Y., Che, W.: Data augmentation approaches in natural language processing: A survey. AI Open **3**, 71–90 (2022)

8. Liu, P., Wang, X., Xiang, C., Meng, W.: A survey of text data augmentation. In: 2020 International Conference on Computer Communication and Network Security (CCNS), pp. 191–195. IEEE (2020)

9. Queiroz Abonizio, H., Barbon Junior, S.: Pre-trained data augmentation for text classification. In: Cerri, R., Prati, R.C. (eds.) BRACIS 2020. LNCS (LNAI), vol. 12319, pp. 551–565. Springer, Cham (2020). https://doi.org/10.1007/978-3-030-61377-8_38

10. Sabty, C., Omar, I., Wasfalla, F., Islam, M., Abdennadher, S.: Data augmentation techniques on Arabic data for named entity recognition. Procedia Comput. Sci. **189**, 292–299 (2021)

11. Siagh, A., Laallam, F.Z., Kazar, O.: Building a multilingual corpus of tweets relating to Algerian higher education. In: International Conference on Intelligent Systems and Pattern Recognition, pp. 132–138. Springer, Cham (2022). https://doi.org/10.1007/978-3-031-08277-1_11

12. Tang, H., Kamei, S., Morimoto, Y.: Data augmentation methods for enhancing robustness in text classification tasks. Algorithms **16**(1), 59 (2023)

13. Wei, J., Zou, K.: Eda: Easy data augmentation techniques for boosting performance on text classification tasks. arXiv preprint arXiv:1901.11196 (2019)

14. Wu, J.L., Huang, S.: Application of generative adversarial networks and shapley algorithm based on easy data augmentation for imbalanced text data. Appl. Sci. **12**(21), 10964 (2022)

15. Wu, X., Lv, S., Zang, L., Han, J., Hu, S.: Conditional BERT contextual augmentation. In: Rodrigues, J.M.F., et al. (eds.) ICCS 2019. LNCS, vol. 11539, pp. 84–95. Springer, Cham (2019). https://doi.org/10.1007/978-3-030-22747-0_7

16. Yoo, K.M., Park, D., Kang, J., Lee, S.W., Park, W.: Gpt3mix: leveraging large-scale language models for text augmentation. arXiv preprint arXiv:2104.08826 (2021)

17. Yu, Y.J., Yoon, S.J., Jun, S.Y., Kim, J.W.: Tabas: text augmentation based on attention score for text classification model. ICT Express **8**(4), 549–554 (2022)

Building Domain Ontologies for Tunisian Dialect: Towards Aspect Sentiment Analysis from Social Media

Mehdi Belguith[1]([✉]), Chafik Aloulou[1], and Bilel Gargouri[2]

[1] ANLP-RG, MIRACL Laboratory, FSEGS, University of Sfax, Sfax, Tunisia
Belguith.mehdi2017@gmail.com, chafik.aloulou@fsegs.usf.tn
[2] MIRACL Laboratory, FSEGS, University of Sfax, Sfax, Tunisia
bilel.gargouri@fsegs.usf.tn

Abstract. Sentiment analysis from social media has received increasing attention during the last few years. Indeed, getting the opinion of customers, users, or people about a product or a service is paramount for many firms in different sectors, ranging from food to politics. Recently, Aspect Sentiment Analysis (ASA) constitutes a rapidly evolving research area. The idea is to determine for each comment, not only an overall polarity (i.e., positive, negative, or neutral) but a detailed polarity per aspect. This paper proposes an original work for building domain ontologies towards aspect sentiment analysis on Tunisian social media. These ontologies will be exploited to improve classification performance for both aspect detection and aspect sentiment analysis. To our best of knowledge, there is no Tunisian ontology extracted from social media content dedicated to sentiment analysis or any other close research domain. Our proposed method is based on an NLP pipeline that we apply on raw Tunisian dialect datasets scraped from social media. We focused on four domains which are Mobile phones, Food, Tunisian election and Radio/TV Programs and applied our proposed method to build Tunisian dialect domain ontologies.

Keywords: Domain ontology · Sentiment analysis · Aspect · NLP pipeline · Tunisian dialect · Social media

1 Introduction

With the rise of social media and the increased number of user generated content, many firms are really interested in extracting the opinion of their customers about the product/service in order to take the right decision at the right moment. However, analyzing the big amount of available information on different social media, is far exceeding human processing capabilities. In fact, the sharing of opinions has become frequent and sentiment analysis really helps organizations understand consumers' opinions about their products/services.

In our work we focus on Aspect Sentiment Analysis (ASA) on social media. The idea is to determine for each comment, not only an overall polarity (i.e., positive, negative, or neutral) but a detailed polarity per aspect. For example, in the comment "

باهية qualité كان حتى غالية ياسر منشرياهش" (I won't buy it, It's too expensive even though it is of good quality), ASA would first detect the aspects which are " سوم" (price) and نوعية/qualité (quality). Then, it determines the sentiment "positive" for "quality" and negative for "price".

Currently, the use of domain ontologies has gained prominence in the field of sentiment analysis and have proven that they are able to improve performance for both aspect detection and sentiment classification [1–3].

Most prior studies on sentiment analysis have focused on Indo-European languages [4, 5], mainly on English, French, etc. However, until the writing of this paper, few research studies using ontologies for sentiment analysis were developed for Arabic [6] and to the best of our knowledge there are no domain ontologies dedicated for aspect sentiment analysis for the Tunisian dialect.

In this research work, we are interested in Tunisian Dialect (TD). We propose an original method for building domain ontologies towards sentiment analysis for TD from social media content.

This paper is structured as follows. In Sect. 2, we present an overview of related work. Then, we briefly describe in Sect. 3, the Tunisian dialect. Then, we propose in Sect. 4 the domain ontology design. Section 5 is dedicated to data scraping from social media. In Sect. 6, we detail our proposed method for building domain ontologies based on an NLP pipeline. The ontology design is then discussed in Sect. 6. Then, in Sect. 7 we present the implementation and discuss obtained results. Finally, we present the conclusion and some interesting perspectives.

2 Overview of Related Work

Although there are a variety of ontologies addressed for some NLP tasks as text classification, there are few ontologies developed for sentiment opinion mining and more precisely for aspect sentiment analysis.

Despite the rare works on ontology driven sentiment analysis, ontologies have proven to be efficient for both tasks of aspect detection and opinion mining. The scarcity of work on ontology-based sentiment analysis could be explained by the fact that most domain sentiment analysis ontologies are developed manually and are time consuming, especially because one has to build an ontology for each domain.

In this section, we first present existing work on Tunisian dialect ontologies and then we focus on related work on ontology-based sentiment analysis. Since there is no related work for Tunisian dialect and very little work for Arabic and its dialects, we present main related works for English since it represents the most used language for research.

In the literature there is only one research work that addressed Tunisian Dialect ontologies. This work is conducted by the ANLP-RG research group, which published the first works related to Tunisian dialect processing. Graja et al. [7] and Karoui et al. [8] proposed a method to understand spoken Tunisian dialect based on a Tunisian dialect ontology. The aim behind building the ontology is to use the ontological concepts for semantic annotation and the ontological relations for speech interpretation. The TD ontology is based on the TuDiCoI Tunisian Dialect Corpus which is a corpus of spoken dialogue related to a specific domain, the railway information service. This ontology does not address sentiment analysis nor social media content.

Some ontologies have been developed for other languages (mainly English and some other European languages) are used to guide sentiment analysis and more precisely Aspect sentiment Analysis. Kontopoulos et al. [9] developed a sentiment analysis ontology towards concept detection. They used data from English tweets dealing with smartphones. The proposed ontology is based on machine learning techniques and contains concepts related by hierarchical relationships. The ontology helps to determine sentiment scores for the different notions in the tweet.

Thakor and Sasi [10] proposed an ontology-driven method for sentiment analysis. The aim is to detect negative posts in social media and particularly in Twitter. The authors collected a big data set of tweets. This data set is analyzed using a dependency parser. Then, they extract the concepts (i.e., the nouns) and the properties (i.e., the verbs). The process of extracting concepts and properties to the ontology has been done manually by the authors. The developed ontology has less than 100 objects and was built based on a small dataset of 250 tweets.

Schouten et al. [5] proposed an aspect-based sentiment analysis method which uses an ontology for the restaurant domain. The ontology has improved the aspect detection and the aspect sentiment analysis.

Zhuang and al. [11] explored the possibility of improving knowledge-driven aspect-based sentiment analysis (ABSA) in terms of efficiency and effectiveness. They proposed a method called SOBA (Semi-automated Ontology Builder for Aspect-based sentiment analysis) to build domain ontologies semi-automatically. Authors consider that building ontologies Semi-automatically could produce more extensive and efficient ontologies on one hand and shorten by 50% the building human time on the other hand. Thus, they created ontologies for the restaurant and laptop domains.

Sharmaand et al. [1] proposed to use domain ontologies for multi-aspect sentiment analysis from hotel reviews. They have improved aspect identification by detecting the hidden aspects in the reviews. They used dependency parsing and neighborhood relations to extract opinion expressions pertaining to the identified aspects. They performed both document (review) level sentiment analysis and aspect level sentiment analysis using three different machine learning techniques (SVM, Naive Bayes and MaxEnt). Best results are achieved by the SVM classifier.

It is worthwhile to conclude that the design of the ontology depends on the objective behind it and the application that will use it. For aspect sentiment analysis, it is more interesting to use domain ontologies, since they could guide the process of aspect detection and opinion classification. Even though we are not able to construct an ontology that englobes all domains, still we can propose a method to build a domain ontology from social media content. We consider that social media content is the most natural and real data source to build domain ontologies for sentiment analysis of social media.

Moreover, according to the presented state-of-the-art, it seems that it is more interesting to build ontologies in a semi-automatic way in order to achieve comparable quality as that of manually built ontologies and also to gain human time and effort (compared to manually built ontologies). Thus, we propose in this paper a method to build Tunisian dialect ontologies in a semi-automatic way and based on social media content.

3   A Brief Presentation of Tunisian Dialect

Tunisian Dialect is considered as a subset of the Arabic dialects of the Western group and belongs to the Maghrebi dialects. TD is also known as "al-Tounsi" or "al-Darija".

Tunisians speak the Tunisian Dialect (TD), a dialect whose vocabulary is mainly Arabic but also contains many non-Arabic words such as Amazigh, Turkish, Italian and, to a large extent, French, due to historical considerations dating back to the 18th and 19th centuries. However, Arabic is considered as the official language in Tunisia even though Tunisian people speak the Tunisian dialect. Indeed, TD is rarely used in written books or in official ceremonies or Tv programs.

Tunisian dialect differs from Modern Standard Arabic MSA in different levels [12]. At the lexical level, as we explained above, TD vocabulary is not only composed of Arabic words, but it also uses many French, Berber, Italian, Spanish words. Moreover, Unlike MSA, most TD content does not respect any orthographic standard even though a conventional orthography was proposed by [13].

At the morphological level, there are great differences between MSA and TD. We can note the absence of some verb modes of as well as the form of the duel. Unlike MSA, verb conjugation in DT is the same for the first person and second person in singular and do not distinguish between the feminine and masculine, and do not use any specific conjugation for the dual [14].

4   TSA Ontology Design

To design our Tunisian Sentiment Analysis (TSA) ontology, we have to answer three main questions that guide the building of a domain ontology. First, we have to determine the entities that should figure in the domain ontology dedicated for opinion mining in the specific domain. Second, we have to find a way to group these entities, in a top-down hierarchy. And third, we have to think about a methodology to model the entities in order to easily support the task of annotating data towards sentiment analysis.

Our methodology for modeling the entities of a domain ontology consists of two main steps. In the first step, we determine the vocabulary (terminology) that is most used in the considered domain. For that, we use the scraped data from social media. Note that for each considered domain we have the corresponding data Set. Second, we define the entities (i.e., concepts, properties, etc.) that are useful for sentiment analysis annotation tasks which consist of determining the sentiment value of the whole comment (i.e., positive, negative or neutral) as well as the polarity (sentiment value) for each aspect that appears in the comment.

4.1   Ontology Entities

For our TSA ontology, we propose to consider three main entities that are the most useful for the sentiment analysis annotation task. More precisely, we consider three entities which are the Sentiment class, the Target class and the SentimentExpression class.

The Sentiment Class. Represents the different polarities or values of a sentiment. In our case, we are interested in three polarities which are positive, neutral and negative.

The Target Class. Represents the aspects. Note that we define a top-down hierarchy for the different aspects according to the relationship between them. The number of classes and subclasses are not defined in advance since they are extracted from data that represents each specific domain. The higher-level concepts correspond to aspect categories, while the subclasses are often target expressions of an aspect. For example, for the "restaurant" domain, the Target Subclasses are Service, Location, cleanliness, etc.

The target class can have one or multiple aspects linked in a top-down hierarchy. For example, Food aspect has many sub-aspects such as price, quality, etc. which we denote by FOOD#PRICES, FOOD#QUALITY, (the aspect, #the sub-aspect). We can have more the two hierarchies as for MOBILE-PHONE#BATTERY#CAPACITY, RESTAURANT#FOOD#PIZZA#PRICE, etc.

The SentimentExpression Class. Represents the vocabulary of sentiment words used to express the sentiment as for example "good", "bad", "tasty", "expensive". etc. Note that each sentiment expression is related, by *hasSentiment* relation, to a Sentiment value (i.e., Positive, negative or neutral). It is also related to a Target, with the *hasTarget* relation. Generally, the *hasTarget* relation points to the top-level concept Target. So, for example, the word "good" has a positive sentiment value for all targets. However, for a few cases, this is not true since for example the word "cold" has a positive sentiment value when it is linked with "Lemonade" and a negative one when it refers to "Soup".

This ontology design formed of three entities described above, allows us to perform the two main tasks of ASA which are aspect detection and sentiment classification.

For aspect detection, when for example we encounter, in a comment, a sentiment word such as "tasty", we know that its target is something related to the aspects "food" or "drink" even though the comment does not explicitly contain these words. Also, when we encounter the word "expensive", we know that it refers to the product/service "price".

For sentiment classification, when we encounter a sentiment word in a comment, the aspect for which we want to determine the sentiment value (positive, negative or neutral) has to be of the same type as the target of that word. Otherwise, the sentiment word will not be considered for that aspect. For example, when we want to determine the sentiment value of the aspect Food-Price and we encounter the word "tasty" we should not take the positive value of "tasty" into account, since it is not relevant to food-price. This is especially useful when a comment has more than one aspect. For example, in the "expensive but tasty", the aspect FOOD#PRICE will have a negative value however the FOOD#QUALITY will have a positive sentiment value.

4.2 Ontology Semantic Relations

We propose the following semantic relations for TSA ontology:

- **has-a-polarity**: this relation links the **SentimentExpression class** to the **Sentiment class.** So for example, the word "good" is related to the positive polarity and the word "cold" could be related to "positive" or "negative polarity according to its **Target class** (or subclass). So, if "cold" is related to the target "Limonade", it has a positive polarity and if it is linked to "Pizza" for example it has a negative polarity.

- **has-aspect:** this relation links the **Target class** with its aspects. For example, the target class "Restaurant" is related by the "has-aspect" relation to the aspects "Location", "food", "service", etc.

- **subclass-of:** this relation links the aspects to its sub-aspects. For example, the aspect "food" has many subclasses (Pizza, burger, etc.) that are related to it by the relation "subclass-of".

- **is-a**: This semantic relation represents the relation of synonymy and could be used between entities that belong to the same class. For example, "casse-croute" is-a "sandwich".

5 Social Media Scraping Towards Collecting Data

In order to apply our proposed method for building domain ontologies, we have first to scrape data from social media. Data Scraping (also called Web scraping) is used to automatically collect large data from websites. In our case, we are interested in social media scraping which consists of automatically extracting comments from social media such as Facebook, twitter, Instagram, YouTube, etc. for sentiment analysis purposes.

So, the main idea is to extract from unstructured data (social media in our case), structured data (the comments or the opinions) (see Fig. 1).

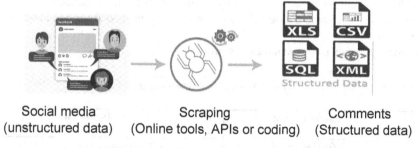

| Social media
(unstructured data) | Scraping
(Online tools, APIs or coding) | Comments
(Structured data) |

Fig. 1. Data Scraping from social media

There are different ways for data scraping such as online tools or services (e.g., Instant data Scraper, Web scraper, BeautifulSoup, selenium, Facebook scraper, youtube_comment_scraper), APIs, or coding.

To scrap our data, we used Python since it is more suitable for web scraping and allows many libraries that facilitate the scraping ("Selenium" for web testing and browser activities, "BeautifulSoup" for parsing HTML and XML documents, "Pandas" for data extraction, manipulation and analysis).

We collected a first dataset called YCTSA (YouTube Corpus for Tunisian Sentiment Analysis). YCTSA refers to four domains (Mobile phones, Food, Tunisian election and TV Programs) and it contains 23275 Tunisian comments written in Tunisian Arabic, Tunisian Arabizi or both.

We also used two other known datasets that are also collected using Web scraping techniques (Table 1):

- **MHB Dataset** collected by [15] from "Facebook" and related to the supermarket domain and mainly to Tunisian 5 supermarkets (i.e., "Magasin general", "Monoprix", "Carrefour", "Aziza" and "Geant"). This dataset is collected by using two tools (Facepager[1] and Export Comments[2]).
- **The TSAC Dataset**[3] collected by [16] from official pages of Tunisian radios and television channels mainly "Mosaique FM", "Jawhara FM", "Shams FM", "Hiwar Ettounsi TV" and "Nessma TV".

Table 1. Size of the datasets in terms of comments, words and unique words

Dataset	# Comments	# Words	# Vocabulary
YCTSA	23 275	273 979	39 487
MHB	17 810	127 990	29 995
TSAC	17 060	112 596	42 129
Total	58145	514 565	111 611

Based on the three presented datasets, we extracted four domain datasets referring to the following domains: Mobile phones, Food, Tunisian election and Radio/TV Programs. Note that we combine together comments referring to the same domain. For example, we collected in the same dataset, all comments related to food from the YCTSA and MHB datasets.

Table 2 presents some examples of comments.

Table 2. Examples of comments

Raw comment	Translation
حوار طاااير بصراحة وهاك الصحافي بالحق محترف وعندو ما يقول بروحه	Excellent dialogue honestly ... and the journalist is really professional and has many things to say
ياسر بنينة البيتزا زادة اما غالية ب دينار 30 شفم ياخي	The Pizza is very delicious but it is expensive... 30 dinars what's happening

Table 3 presents the number of comments related to the four domain data sets. Note that we tried to use datasets that are almost balanced to avoid biased results especially for sentiment analysis.

---

[1] https://github.com/strohne/Facepager.

[2] https://exportcomments.com/.

[3] https://github.com/fbougares/TSAC.

Table 3. Number of comments for the 4 retained domains

Domain	# Comments
Mobile phones	10931
Food	9229
Tunisian election	8183
Radio/TV Channels	11992
Total	40335

One note that we apply on these domain datasets, the NLP pipeline proposed in Sect. 6 in order to construct the lexicon (vocabulary) that will be used by huma for building the domain ontologies.

6 Proposed NLP Pipeline

We propose in this section, our method for building a TD ontology for sentiment analysis. As shown in Fig. 2, our method relies on an NLP pipeline for raw dataset processing. The input of the pipeline is a raw dataset (i.e., in our case a list of comments for each domain) and the output is a domain vocabulary (i.e., a list of words list of Our proposes NLP pipeline is composed of 11 steps: Snippet splitting, Data cleaning, Transliteration, Orthographic normalization, Spelling error correction, Emoji conversion into textual meaning, Foreign word translation, Redundant letters processing, Light stemming and POS tagging.

Splitting. This step consists in splitting the comments (in the form of snippets) into tokens. The splitting relies on the space character as well as the punctuation marks when available.

Cleaning. This step consists in deleting diacritics, numbers, unuseful signs, punctuation marks, URLs, etc. that have no interest in the ontology building nor in the opinion mining.

Transliteration. Consists of converting comments written with Tunisian Arabizi (i.e., using Latin letters and numbers) into comments written with Tunisian Arabic (i.e., using Arabic letters only). Note that Tunisian young people may use Latin characters to refer to Arabic letters (that have almost the same pronunciation as the Latin ones, such as " ب"/"b"). They also may use numbers to replace letters that **do** not have an equivalent (in terms of pronunciation) in the Latin alphabet. For example, they use the number "3" to replace the letter " ع" since they are graphically very close.

Table 4 presents a comment from our dataset on which we apply the transliteration pre-processing task.

Orthographic Normalization. Since social network content does not follow any spelling convention, we propose in this step to normalize the letters that could be written in different ways (such as the letter " أ" "alif" which can be written in different ways: " أ", " إ", etc.).

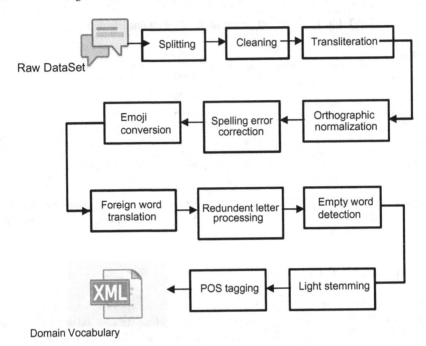

Fig. 2. NLP Pipeline for raw dataset processing

Table 4. Example of a comment before and after transliteration

	Comment
Before transliteration	Alndhafh w elbena hiya elkol ya3tikom saht
After transliteration	النظافه و البنا هي الكل يعطيكم الف صحة
English Translation	The cleanliness and the good taste are the main key may God give you good health

We apply the orthographic normalization system [17] based on the orthographic convention CODA-TUN [18].

Spelling Error Correction. In this step, we also apply the orthographic normalization system [17] to correct spelling errors in the datasets. For instance, the word "sokor" is replaced by "sukkar" (meaning "sugar").

Emoji Conversion into Textual Meaning. Emoji are often used in social media to enrich users' emotions. They play an important role in the task of social media sentiment analysis. In practice, many researchers do not consider them and delete them from the comments [19]. However, they could enhance results in opinion mining [20, 21]. In this step we propose to convert emoji into their textual meaning. For example, the heart emoji ❤is converted into " حب" (love) which refers to a positive sentiment.

Foreign Word Translation. As we mentioned in Sect. 3, TD contains many foreign words. This step consists in translating foreign words (mainly in French) into Tunisian. Since the foreign words are a normal phenomena in many dialects including TD, we decided not to throw them away since they could be important for any NLP task including ontology construction and sentiment analysis. For example in the comment " قعدت ساعة نستنى باش حضرت Pizza طلبت ", the word "Pizza" will be translated into " بيتزا".

Redundant Letters Processing. Most research works remove duplicate or redundant letters inside the words and consider that, by doing that, they correct the spelling errors. However, in social media content, the redundant letters are used intentionally (not as a mistake) to emphasize on the word or express the meaning of "very"/"so much". Thus, we propose in this step to add the textual meaning of the repeated letters before removing the duplicated letters. So, we add the TD word " برش" which means "very"/"so much" after the considered word. The idea is to keep the real meaning behind the duplicated letters that could enhance the opinion mining task. For example the word " صغييير" (little) is replaced by " صغير برش" (very little).

Empty Word Detection. In most research works related to sentiment analysis, empty words such as coordinating conjunction (e.g., " و" (and)) and pronouns (e.g., انت/انتي(you -singular)), are considered as empty words since they do not affect the opinion detection. In this work, we also delete the empty words since they have no interest for building the ontology nor for sentiment analysis.

Light Stemming. The light stemming consists of eliminating the prefixes and suffixes of the word, so that words with similar meaning will be considered as the same word (concept). Note that the prefixes represent both definite articles and conjunctions. However, the suffixes correspond to the word's endings and indicate the gender, the number or the personal pronoun. For example, the word " الديموقراطية" (the democracy) will be transformed into " ديموقراطية" (the definite article " ال" (the) is removed).

Table 5 presents the comment of Table 3, before and after applying the light stemming.

POS Tagging. Part Of Speech (POS) tagging allows to assign for each word its POS (verb, noun, adjective, conjunction, etc.). This step is crucial since it allows us to determine the concepts and the relations of the ontology. Thus, some POS categories such as the nouns will represent the aspects (i.e., The target class) and others like the adjectives will express the sentiments (i.e., the SentimentExpression class). However, the verbs could refer to the relations between the target and its aspects or sub-aspects.

In this work, we propose to use the Tunisian Tagger Social Media Dialect (SMD) [16] for the POS tagging since it is specific to Tunisian dialect used in social media. SMD allows constructing the syntactic tree of the sentence (the comment in our context) in addition of giving the POS of each word in the tree. Figure 3, presents the output of the SMD for the comment " نموت على الاكل التونسي المحرحر" (i.e., I adore Tunisian spicy food).

Table 5. Example of light stemming

	Comment
Before light stemming	النظافه و البنا هي الكل يعطيكم الف صحة
After light stemming	**نظافه و بنا هي كل يعطي الف صحة**
English Translation	**Cleanliness** and **good taste** are **main key** may God give you good health

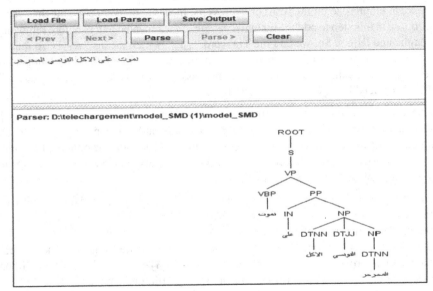

Fig. 3. Example of a comment processed by *SMD Tunisian tagger*[2]

7   Ontology Building

We propose to build our domain ontologies in a semi-manual way, based on the output of the NLP pipeline that we apply on our datasets. Indeed, according to the state-of-the-art presented in Sect. 2, it is more interesting to build ontologies in a semi-automatic for two reasons. The first reason is that compared to manually built ontologies, the semi-automatic way is less consuming human time and effort. The second reason is to achieve comparable quality as that of manually built domain ontologies.

We used the Python programming language to implement our NLP pipeline. As we explained in Sect. 6, the aim of the NLP pipeline is to clean and process the raw datasets composed of Comments scraped from social media. The idea is to facilitate the task of determining the entities of the ontologies (the targets, the aspects and the sub-aspects, the sentiment expressions, the sentiments and the relations).

After applying the proposed NLP pipeline on the datasets, we obtain for each domain (Food, Mobile phone, Election, TV/Radio channels) a set of words with their corresponding POS.

Table 6 presents the numbers of words (before and after the NLP pipeline) classified according to their POS as well as some examples for each category.

The word sets represent the lexicons referring to the considered domains. Each word set is then classified manually, by two experts as follows:

- Targets: words which POS is a noun such as " رستوران/" (restaurant), " تلفون" (mobile phone), " انتخاب" (election), etc.
- Aspects and sub-aspects: words which POS is also a noun such as " بيتزا" (Pizza), " سوم" (Price), " سرعة" (speed), " نظافة" (cleanness), " سرفيس" (service), etc.
- Sentiment expressions: mainly the words which POS is an adjective such as " خايب" (bad), " باهي" (good), " بارد" (cold), " غالي" (expensive), " يفدد" (boring), etc.

Table 6. Number of words before and after the NLP pipeline

Domain	# unique words before NLP pipeline	#words after NLP pipeline	#POS - noun	#POS - adj	#POS - verb
Mobile phones	12863	2860	2305 e.g. سوم (Price)	327 e.g. غالي (expensive)	228 e.g. شريت (I bought)
Food	16604	3270	2524 e.g. بيتزا (Pizza)	314 e.g. بنينة (Tasteful)	432 e.g. طلبت (commended)
Tunisien election	21352	4504	3741 e.g. انتخاب (election)	451 e.g. بكري (early)	312 e.g. ننتخب (I vote)
Radio/TV Programs	22785	4117	3024 e.g. منشط	496 e.g. يفدد (boring)	597 e.g. يقدم (present)

Note that the aim of this work is to build domain ontologies that will be used to enhance the TD aspect sentiment analysis task. Despite the rare works on ontology driven sentiment analysis, they have shown their efficiency for both tasks of aspect detection and aspect sentiment analysis.

Our domain ontologies are composed of three main classes: the sentiment class (positive, neutral and negative), the target class (the aspects, the sub-aspects), the sentiment expression class (the sentiment vocabulary). It also contains four relation types between the classes which are has-a-polarity relation (i.e., links the SentimentExpression class to the Sentiment class), has-aspect relation (i.e., links the Target class with its aspects), subclass-of relation (links the aspects to its sub-aspects) and the is-a relation (i.e., represents the relation of synonymy between entities that belong to the same class).

To build our domain ontologies, we used the "Protégé" open-source platform. To keep our ontologies manageable, we have deliberately opted for a relatively small, but focused, ontologies.

Figure 4 presents an extract of the Tunisian election ontology which shows the target class " انتخاب" (election) and its semantic expressions (elections, elect, choose, choice, etc.). Since the presence of one or more of these expressions in the comment refers to the target, they are related to it by the semantic relation (is-a/a-sort-of).

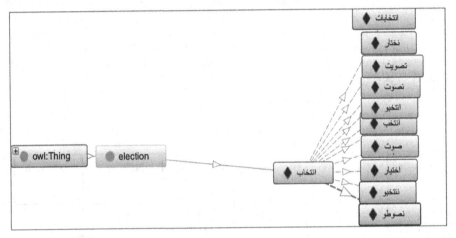

Fig. 4. Extract from the Election domain Ontology

As we can notice the target class " انتخاب" (election) has relations (i.e. a-sort-of/is-a) with 10 other aspects (or semantic subclass) which are انتخابات(elections), نختار(choose), تصويت(vote), نصوت(I vote), انتخبو(elect), انتخب(I elect), صَوّت(vote), اختيار(choice), ننتخبو(we elect), نصوطو(we vote).

Table 7 presents an example of raw comments extracted from our raw dataset (Mobile phone dataset). The second row shows the result of applying the NLP Pipeline (the 11 NLP steps) on the comment. The third row represents the expert annotations. So based on the list of domain words and their POS determined by the NLP pipeline, our experts annotate first the target (generally, with the domain of the dataset and in our example the domain is " تلفون"/MOBILE PHONE). Then, they annotate the aspects and the sub-aspects related to the Target. Generally, they have a POS of a noun and in our example, there are two aspects: the aspect " تصميم"/design and the aspect سرعة/Speed which is deduced implicitly from the adjective رزين/slow). They also annotate the sentiment expressions which generally represent the adjectives related to the domain (in our example there are two adjectives which are: ناقص/Uncomplete) and رزين/slow).

Moreover, they annotate six semantic relations as follows:

– Two semantic relations between the target and their aspects: has-aspect (MOBILE PHONE, DESIGN) and has-aspect (MOBILE PHONE, SPEED).

- Two semantic relations between the aspects and the sentiment expressions describing them: has-an-expression (SPEED, LOW) and has-an-expression (DESIGN, UNCOMPLETE)
- Two semantic relations between the sentiment expressions and their sentiment: has-a-polarity (Uncomplete, NEGATIVE) and has-a-polarity (slow, NEGATIVE).

Table 7. An example of expert annotation of a comment

Raw Comment	tasmim telifoun na3is w t7issou rzin
After NLP Pipeline: List of the domain words + POS	" تلفون" (Mobile phone): nouns; "تصميم" (Design) : noun "ناقص"(Uncomplete): adjective; "رزين" (slow): adjective
Expert annotation based on the NLP pipeline output	Target: PHONE Aspects: * تصميم/design + * Speed/سرعة Negative Sentiment expressions: *ناقص (Uncomplete) + * رزين)slow)

Table 8 presents for each domain ontology, the number of sentiment expressions, target concepts and the maximum depth of its class hierarchy. Note that for all ontologies we used three sentiment classes: positive, neutral and negative.

Table 8. Number of words before and after the NLP pipeline

Domain ontology	#Sentiment expression	#Target concept	#Depth hierarchy
Mobile phones	67	221	4
Food	56	185	5
Tunisian election	78	171	3
TV Programs	83	235	3

8 Conclusion and Perspectives

In this research work, we have proposed a method to build domain ontologies for Tunisian Dialect in order to enhance the aspect sentiment analysis task. To the best of our knowledge there is no such ontologies for TD. Our proposed method is based on an NLP pipeline and takes into account the specificities of the Tunisian dialect. The pipeline is composed of many NLP steps: Snippet splitting, cleaning, Transliteration, Orthographic normalization, Spelling error correction, Emoji conversion into textual meaning, foreign word translation, Redundant letters processing, Light stemming and POS tagging.

The proposed method is implemented and applied on three datasets scraped from social media: YCTSA dataset, the MHB Dataset and the TSAC Dataset. From these datasets, we constructed four datasets which refer to four domains (Mobile phones, Food, Tunisian election and TV Programs).

Our proposed method to build domain ontologies is semi-automatic. It is based on the output of the NLP pipeline that we applied on the domain datasets. The semi-automatic method, compared to manually built ontologies, allows the gain of human time and effort and achieves comparable quality as that of manually built domain ontologies.

Our domain ontologies are composed of three main classes: the sentiment class (positive, neutral and negative), the target class (the aspects and the sub-aspects) and the sentiment expression class (the sentiment vocabulary). The classes are linked by different semantic relations such as has-a-polarity, has-aspect, has-sub-aspect, is-a, etc.).

It is important to notice that one should handle the negation phenomenon which differs from one Arabic dialect to another and from dialects to Modern Standard Arabic [6]. Indeed, we have to propose a solution to detect the negation in the comments. Thus, in the example " البيتزا هذي موش بنينه " (This pizza is not delicious), we have to consider the negation particle " موش " (is not) otherwise the comment will be assigned a positive opinion instead of a negative one.

As future perspectives, we would like to study the effect of using our TD domain ontologies on sentiment analysis from social media. We also intend to exploit these ontologies to automatically detect the aspects in the comments.

References

1. Sharma, S., Saraswat, M., Dubey, A.K.: Multi-aspect sentiment analysis using domain ontologies. In: Villazón-Terrazas, B., Ortiz-Rodriguez, F., Tiwari, S., Sicilia, M.A., Martín-Moncunill, D. (eds.) Knowledge Graphs and Semantic Web, KGSWC 2022. CCIS, vol. 1686, pp. 263–276. Springer, Cham (2022). https://doi.org/10.1007/978-3-031-21422-6_19

2. Ten Haaf, F., et al.: WEB-SOBA: word embeddings-based semi-automatic ontology building for aspect-based sentiment classification. In: Verborgh, R., et al. (eds.) The Semantic Web (ESWC 2021). LNCS, vol. 12731, pp. 340–355. Springer, Cham (2021). https://doi.org/10.1007/978-3-030-77385-4_20

3. Dera, E., Frasincar, F., Schouten, K., Zhuang, L.: SASOBUS: semi-automatic sentiment domain ontology building using synsets. In: Harth, A., et al. (eds.) The Semantic Web ESWC 2020. LNCS, vol. 12123, pp. 105–120. Springer, Cham (2020). https://doi.org/10.1007/978-3-030-49461-2_7

4. García-Díaz, J.A., Cánovas-García, M., Valencia-García, R.: Ontology-driven aspect-based sentiment analysis classification: an infodemiological case study regarding infectious diseases in Latin America. Futur. Gener. Comput. Syst.. Gener. Comput. Syst. 112, 641–657 (2020)

5. Schouten, K., Frasincar, F., De Jong, F.: Ontology-enhanced aspect-based sentiment analysis. In: Cabot, J., De Virgilio, R., Torlone, R. (eds.) Web Engineering ICWE 2017. LNCS, vol. 10360, pp. 302–320. Springer, Cham (2017). https://doi.org/10.1007/978-3-319-60131-1_17

6. Bensoltane, R., Zaki, T.: Aspect-based sentiment analysis: an overview in the use of Arabic language. Artif. Intell. Rev.. Intell. Rev. 56, 2325–2363 (2023). https://doi.org/10.1007/s10462-022-10215-3

7. Graja, M., Jaoua, M., Hadrich Belguith, L.: Building Ontologies to Understand Spoken Tunisian Dialect. CoRR abs/1109.0624 (2011)

8. Karoui, J., Graja, M., Boudabous, M., Belguith, L.H.: Semi-automatic domain ontology construction from spoken corpus in Tunisian dialect: railway request information. Int. J. Recent Contributions Eng. Sci. IT 1(1), 35–38 (2013)
9. Kontopoulos, E., Berberidis, C., Dergiades, T., Bassiliades, N.: Ontology-based sentiment analysis of twitter posts. Expert Syst. Appl. 40, 4065–4074 (2013). https://doi.org/10.1016/j. eswa.2013.01.001. https://www.sciencedirect.com/science/article/pii/S0957417413000043
10. Thakor, P., Sasi, S.: Ontology-based sentiment analysis process for social media content. Procedia Comput. Sci. 53, 199–207 (2015)
11. Zhuang, L., Schouten, K., Frasincar, F.: SOBA: semi-automated ontology builder for aspect-based sentiment analysis. J. Web Semant. 60 (2019). https://doi.org/10.1016/j.websem.2019. 100544
12. Belguith, M., Azaiez, N., Aloulou, C., Gargouri, B.: Social media sentiment classification for Tunisian dialect: a deep learning approach. In: International Conference on Intelligent Systems and Pattern Recognition (ISPR), Hammamet, Tunisia (2022)
13. Zribi, I., Boujelbane, R., Masmoudi, A., Ellouze, M., Hadrich Belguith, L., Habash, H.: A Conventional Orthography for Tunisian Arabic. LREC (2014)
14. Mekki, A., Zribi, I., Ellouze, M., Hadrich Belguith, L.: Syntactic analysis of the Tunisian Arabic. In: International Conference on Language Processing and Knowledge Management (LPKM), Kerkennah, Tunisia (2017)
15. Masmoudi, A., Hamdi, J., Hadrich Belguith, L.: Deep learning for sentiment analysis of Tunisian dialect. CyS 25, 129–148 (2021). https://doi.org/10.13053/cys-25-1-3472
16. Medhaffar, S., Bougares, F., Estève, Y., Hadrich-Belguith, L.: Sentiment analysis of tunisian dialects: linguistic resources and experiments. In: Proceedings of the Third Arabic Natural Language Processing Workshop, pp. 55–61. Association for Computational Linguistics, Valencia, Spain (2017). https://doi.org/10.18653/v1/W17-1307
17. Besdouri, F., Mekki, A., Zribi, I., Ellouze, M.: Improvement of the COTA-orthography system through language modeling. In: AICCSA (2021)
18. Mekki, A., Zribi, I., Ellouze, M., Hadrich Belguith, L.: Treebank creation and parser generation for Tunisian social media text. In: The 17th ACS/IEEE International Conference on Computer Systems and Applications (AICCSA), Antalya, Turkey (2020)
19. Wunderlich, F., Memmert, D.: Innovative approaches in sports science—lexicon-based sentiment analysis as a tool to analyze sports-related twitter communication. Appl. Sci. 10, 431 (2020). https://doi.org/10.3390/app10020431
20. Nisar, M.A., Hussain, M., Amin, F.: Sentiment analysis on emoticons and emoji using deep learning techniques. Int. J. Adv. Comput. Sci. Appl. (2021)
21. Chen, Z., et al.: Emoji-powered sentiment and emotion detection from software developers communication data. ACM Trans. Softw. Eng. Methodol. 30(2), 18:1–18:48 (2021)

Log Analysis for Feature Engineering and Application of a Boosting Algorithm to Detect Insider Threats

Samiha Besnaci[✉], Mohamed Hafidi, and Mahnane Lamia

LRS Laboratory, computer science department, Badji Mokhtar University, Annaba, Algeria
besnacisamiha@yahoo.fr

Abstract. The insider threat has captured the attention of a large number of researchers, as a sensitive and critical issue for most organizations in today's digital world. It is also a major source of information security and can cause more damage and financial loss than any other threat. In this article, we've used feature engineering for features that represent users' day-to-day activities. We tried different machine learning models such as random forest, xgboost and Catboost. Since the data used to detect malicious activity is unbalanced, the target audience is small. We used KMeansSmote to balance the classes of learning so that the algorithms can learn both classes well. And we used the catboost algorithm to identify the malicious user. The dataset used to evaluate this model is Cert v4.2. CatBoost outperformed other models with the highest F1-score of 95%.

Keywords: Boosting learning · Insider Threat · Machine Learning · Enterprise security · KMeansSmote · CatBoost

1 Introduction

When it comes to the security of an organization's assets–both physical and digital–serious steps are taken to protect them from outside threats. The focus is usually on external threats since they are expected. Organizations are more prone to exploitation by outside malicious actors for a variety of reasons. An example would be the theft of the confidential information they carry, by their competitors for example. However, when it comes to an insider threat, it's usually unexpected and much harder to defend against. One of the most prevalent cybersecurity threats is the insider threat. It is carried out by employees associated with the organization and trusted. This makes it easier for them to stage attacks for their possession of sensitive resources and information. Insider attacks have many causes, such as disagreements with colleagues or bosses, unintentional human error, and pressure from competing organizations. Insider threats target data mining activities to infiltrate and sabotage computer systems. Several recent literatures have dealt with the aspect of insider threats, their spread and their damage. But to this day it does [1]. Nowadays, many investigations show how dangerous the threat is from within. For example, Cyber Security Insiders 2021 Insider Threat Report indicates that 98% of organizations feel vulnerable to insider attacks [2], and malicious insiders cause significant damage and loss to organizations.

A. Bennour et al. (Eds.): ISPR 2023, CCIS 1940, pp. 268–284, 2024.
https://doi.org/10.1007/978-3-031-46335-8_21

In its 2019 report, the Data Breach Investigation (DBIR) [3] notes an increase in internal data breach threats. Insiders estimated global breaches in 2018 at around 28%, and in 2019 they continued to rise to 34%. According to reports from the Ponemon Institute [4], the total average cost of a threat increased by 31% between 2017 and 2019. This averages: $8.76 million in 2017 and $11.45 million in 2019. It is therefore extremely important to identify threats with the lowest possible false positive rate. 206 days is the life cycle needed to identify a breach, which is more than 6 months, and an additional 73 days to repair its damage. This information was provided by the 2019 study [5] on the cost of data breaches. The techniques used by researchers to uncover security issues and mitigate their impact are data analytics. To solve cybersecurity use cases such as malicious activity detection, several techniques such as natural language processing, statistical learning, artificial intelligence, and machine learning are used. Technologies based on data collection, analysis and threat detection [6]. The methods used to detect malicious activity rely on feature engineering and learning representations of malicious user behavior and insider identity threats. These are effective detection methods. Since insider threat data is out of balance. The minority class representing malicious user behavior is far below the normal user class that represents the most [7]. Machine learning algorithms give high accuracy with many samples, while low accuracy with few samples. Imbalanced data is therefore one of the challenges faced by detection systems, and the important category for detection is the small category [8, 9].

Therefore, the main purpose of this article is an in-depth study that aims to find out the insider threat and study how to deal with different data. Every day, the user processes multiple data to complete their work, and the organization's protection system needs to monitor the user's movements by analyzing the huge amount of data. There must be systems that can quickly analyze data and detect insider threats. Therefore, the data should be used to extract the behavioral characteristics of the user and use it through machine learning techniques to learn the normal behavior of each user and to declare any difference in the normal behavior as harmful.

The problem with the data used to find harmful activity is the imbalance where the harmful category represents a very small category compared to the benign category which represents the category that is abundant. In this article, we propose to solve these problems first, we analyze and clean the data and extract the appropriate behavioral characteristics, then use a model that works in general in two stages based on the kMeansSmote technique to augment the data and also aims to eliminate intralayer aberrations while avoiding the generation of noisy samples and imbalances between layers. In the second step, we use the Catboost classifier to learn normal behavior and detect any abnormal behavior. The dataset on which to test this model is v4.2 are data intended to detect insider threats.

In summary, the main contributions discussed in this article are:

- Data processing, calculation of missing values, removal of redundancies, extraction of properties and formation of the natural behavior of the user.
- We analyze the problem of the insider threat detection domain, which is the imbalance of data categories, and we propose to increase the small category, and we see if the results are improved and the insider threats are detected.

- improve the random oversampling by the KMeansSMOTE method which aims to fight the weaknesses of the classification algorithm.
- Using the CatBoost supervised learning algorithm to detect insider threats.
- Comparison of the proposed model with other models based on supervised learning techniques, results showed high accuracy.

The parts of this article are organized as follows. The literature related to our subject is discussed in the second section, the third section explains the data used and we explain all the steps of the proposed model. The fourth section discusses the presented results, and finally the fifth section is the conclusion of the article.

2   Related Work

The problem of insider threats is one of the difficult challenges facing researchers.

The method proposed by [10] tackles the insider threat detection problem. Insider attackers cause significant damage to the organization ranging from $100,000 to $500,000 as they can breach all layers of security and steal sensitive data. Their proposed solution uses the LSTM autoencoder. It was trained and evaluated on the CMU CERT v4.2 dataset which contains 930 normal users and 70 malicious insiders and multiple instances for attacker scenarios. The model achieves accuracy (90.60%), precision (97%) and F1-Score (94%).

AlSlaiman and al. [11] To identify insider threats, a system uses LSTM to save time-based activity of each user based on users' document access behavior, and also sentiment analysis to decrease false positive rate and false rate negative the value of AUC = 97%. Shuhan Yuan and Xintao Wu in [12], Show the relationship between insider threats and deep learning, and the trends and challenges it faces. Raval et al. [13] mentioned malicious user detection using learning techniques machine and some case studies. Fangfang Yuan in [14] introduced the use of deep neural network (DNN) in the detection of malicious activity. It is based on identifying user behavior using the LSTMCNN dual model to find abnormal cases and abnormal activities. This model gave a high result of AUC = 0.9449.

Lindauer in [15] Detects whether a session is legitimate or malicious, extracts features from each user's sessions, and converts them into feature sequences of equal length. TF-IDF representations are used. Feature sequences are fed into the ensemble detector with the LSTM grid as the basis. An ablation study was performed to assess the importance of each component used in the methodology. The results of this approach were also compared to previous results in the literature by evaluating the method on two different versions of the dataset CERT4.2 and CERT 6.2. This method achieved the highest areas under the curve (AUC) in both data sets −99.2% and 95.3%, respectively. Liu et al. [16] Web browsing and email content are used to build a user's psychological profile. Insider threats can therefore be predicted based on these analyzes and proactively detect malicious insiders with negative sentiment. In this article, an effective approach is proposed to proactively detect insider threats based on sentiment analysis of web pages and emails. CMU CERT and Enron Email datasets were used to evaluate the performance of this system.

Researchers have proposed several models to address the problem of harmful insider activities within organizations. One of the challenges they face is the imbalance of data that hinders model training and detection accuracy, because the few category is the category that needs to be detected. There are techniques that increase the proportion of small samples. ADASYN [17] is one of the methods used to generate adaptive synthetic samples. There is also a GAN that generates real samples to solve the imbalance problem [18], which gave great accuracy in identifying insider threats.

Mohammad and all [19], use the Light Gradient Boosting Machine (LightGBM) method to detect abnormal events a simple method that gave high accuracy.

In this article, we propose the KS-Catboost model, which works to detect and identify insider threats within organizations. It is based on Nority Oversampling Technology (KMeansSMOTE) algorithm [20] which generates samples to solve the imbalance problem so that the classifier can identify the malicious user.

3   Freamework

As shown in Fig. 1, the proposed model for solving the insider threat problem consists of several steps, the most important of which are: the first step, which is based on data processing and analysis, and the extraction user functionality, which is considered as the starting point for the second stage, which is the learning stage and the detection stage in which user behavior patterns are used to form an image of the normal user. Any deviation from normal behavior is identified in the detection process as an anomaly and should be reported.

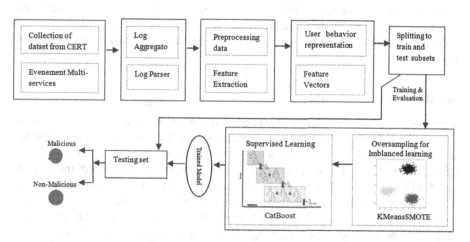

Fig. 1. The proposed general model

In the second section, we noted several solutions that used machine learning and deep learning to find effective solutions to detect malicious activities in organizations. Machine learning (ML) technology is a subset of artificial intelligence that has given machines the ability to perform complex tasks without direct instructions. It performs several tasks including prediction, classification, clustering, and pattern recognition. The training samples that are described by the features are fed into the algorithm to be trained on the inputs. And the resulting information is used to determine if the model has learned well and is able to Three main classifications [21] in which learning algorithms can be grouped:

A. Supervised Learning

In this type, the value of the output is known in advance by the availability of a set of parameterized data and there is a relationship between the inputs and outputs of the learning model. The most common in machine learning and deep learning is supervised learning and it has several applications and most problems are solved with supervised learning. It is classified into two categories, "Classification" and "Regression".

Classification: The outputs of the classification problem are distinct and depend on a specific category. Features found in input samples are grouped into separate classes.

Regression: In the regression problem, the results are presented continuously. To define the input characteristic, continuous functions are used.

B. Unsupervised Learning

Data of this type is unlabeled, where unsupervised learning deals with problems that have an unlabeled data set and the end result is unknown. It derives the structure-based relationship between variables using clustering.

C. Reinforcement Learning

In this type of learning, the algorithms rely on many tuning parameters to predict the outcome of a problem. In this case, the output results are used as input parameters and so on until the ideal output is found. Artificial neural networks and deep learning use this type.

3.1   Data Set

In order to simulate the inner workings of an organization as realistically as possible without violating the privacy and confidentiality of real users of a real organization, the dataset used for such a system should be synthetic data generated by templates to cover different facets of digital organization activity. For the implementation of this system, the CERT dataset [22] was used, especially version 4.2 of the dataset due to the presence of many cases of malicious activity as well as several cases of user activity compared to previous versions. CERT is a collection of synthetically generated data describing several insider threat scenarios in an organization. It includes several types of log files that simulate digital organizational activities over an 18-month period.

Data Generation. To realistically generate the dataset, different aspects of an organization's topology should be considered, such as the social aspect, the behavioral aspect, the preferences of the employees and the content they access, and their ownership of the organization's digital assets [15]. The following models were used for this complex generation of synthetic data:

- **Relational graph model:** which is a representation of organizational relationships between employees. This depicts the social and functional structure of the organization.
- **Asset chart template:** which is a model that describes the ownership of physical and digital assets by employees of the organization, such as PCs, external devices and files.
- **Behavioral model:** a model that describes how each employee typically uses their assets.
- **Communication model:** which depicts the relationship between the employees through the emails sent.
- **Topic model:** describing the user's interests in terms of web browsing trends and email content [15].

Description of Data. The dataset contains 5 different types of log files and their details in Fig. 2:

- **Logon Logs:** contains logs of user login and log out activities.
- **Device logs:** contains logs of user usage of external USB drives.
- **File Logs:** contains logs about file operations performed by the user.
- **HTTP logs:** contains logs of users' web browsing activities.
- **Email Logs:** contains logs of emails sent and received by users.

The dataset also contains an LDAP directory that contains 18 files containing information about employees, their supervisors, business units, and departments within the organization over the 18 months of activity present in the dataset.

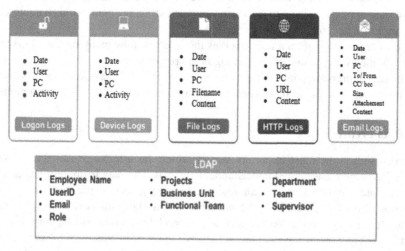

Fig. 2. Dataset Description

Scenarios. Insiders have access to organizational assets, both physical and digital. They take advantage of this by performing malicious scenarios with the intention of stealing confidential information or sabotaging when they feel unhappy or are about to leave the organization. The following 3 scenarios are executed by malicious users and their log traces are present in the dataset:

1. The first scenario simulates an employee who is considering leaving the organization. This employee begins to change his usual behavior from being active only during working hours and not using removable disks to logging on after hours and using removable disks and downloading sensitive data from wikileaks.org.
2. The second scenario simulates an employee looking for a job in a competitor's organization. This can be seen in their web visits to job websites. Right before they leave the company, their USB drive usage increases dramatically and they steal organizational data.
3. Scenario three simulates a disgruntled system administrator who is about to leave the company. This employee uses a USB key to transfer a keylogger to his supervisor's PC and steal his credentials. The system administrator then uses the credentials to send mass emails to the entire organization, causing panic [22].

4 Model Proposed

In this section, we provide details on the proposed KS-Catboost model for detecting insider threats that depend on anomaly detection, as well as the treatment of imbalance in the two data categories.

Before talking about the learning stage, there is a preliminary stage, which is data processing, analysis and extraction of features that give the image of a normal user. This step is considered the most important and also controls the result of the model.

In fact, the data comes from several sources that are collected at the expense of the time of their appearance, and to use them we must extract the behavioral characteristics of the user that are content and time dependent.

For example, how often a user browses the Internet, how many files he downloads per day or the average number of working hours per user.

The step of preparing data, extracting and selecting user behavioral characteristics is the longest step. In order to make each step clearer, we will explain each step separately.

4.1 Data Preprocessing

Pre-processing is the most important step in which the original or raw data is transformed into a meaningful and understandable format. The data is initially inconsistent and incomplete and does not contain features describing specific user behaviors or patterns. And if correctly entered into the machine learning algorithms without modification, it generates several errors, and the accuracy of model detection will be very low. Pretreatment is therefore the best solution to solve these problems. Several steps should be followed, such as cleaning data, removing duplicate values, modifying values, and identifying outliers. When the complete data processing is finished, the characteristics related to each file are extracted.

4.2   Feature Extraction

In this system, we ingest data from 5 different log domains, namely: login/logout, device, file, email and HTTP. We set the working hours from 7:00 a.m. to 6:00 p.m., and to identify each user's PC, we count each user's logins on different PCs, and choose the most frequent as the user's PC, and the others can be shared or belong to other users. This gives us a one-to-one mapping between users and PC which will be used later for feature mining. We then aggregate the daily actions of each domain separately and extract 2 different feature sets adapted from the literature, which we call fine grained and coarse-grained feature sets. We then train the models on each feature set separately, and choose the one that achieves the best performance. In the coarse-grained feature set [23], we count the number of events in each domain like number of logins, number of file operations, number of websites visited, but we don't go into too much detail about the type of file that was open., or what type of website was visited. This gives us 33 features in total, described in Table 1.

Table 1. Coarse-grained feature set.

Domain	Features
Logon	#Logons, #PCs logged on, #after hour logons, #logons on user's PC, #logons on other PC(s)
Device	#device accesses, #PCs with device access, #after hour device accesses, #device accesses on user's PC, #device usage on other PC(s)
File	#file accesses, #PCs with file accesses, #distinct files, #after hour file accesses, #file access on user's PC, #file accesses on other PC(s)
HTTP	#web visits, #PCs with web visits, #URLs visited, #after hour web visits, #URLs visited from other PC(s)
Email Sent/Received	#emails, #distinct recipients, #internal emails, / #internal recipients, #emails sent after hour, #emails with attachment(s), #emails sent from other PC(s)

On the other hand, in the fine-grained feature set [24], we extract more information about each action performed by users. This information is divided into two parts, counting characteristics and statistical characteristics. In counting features, we calculate detailed event frequencies like the number of file operations on different file types (doc, zip, etc.), their sizes, whether performed locally or on a removable drive. In addition to that, we divide websites by subject to have social websites, cloud services, hacktivist websites and job search websites. For each of these events, we count their frequencies on the user's PC, on shared PCs, and whether they occurred during working hours or after working hours. This gives us 360 features.

4.3 Vector Representation of Features

At the end of the data preprocessing phase and the extraction of the characteristics of each file, the data must be labeled as normal or abnormal, for the malicious user it is marked by 1 and for the normal user by 0. The feature vector of each user is obtained for each day and it is the daily behavior pattern of the user. Features contain numeric and textual values that must be processed to avoid skewing the results. In this article, we use MinMaxScaler to limit feature values between $[-1, 1]$ to be compatible with machine learning algorithms. Finally, the data is divided into two groups, the first group assigned to training represents 80% and the second group assigned to testing represents 20%.

4.4 KS-CatBoost Training

In this article, the data used is unbalanced. The rare samples represent the target group for detection, which is the harmful samples, compared to the majority category, which is the normal category. This issue affects the performance of machine learning algorithms that give low accuracy in identifying small samples, leading to losses for organizations due to their inability to give accurate results when ranking. Many techniques are used for sampling or desampling. Subsampling is the selection of a random subset of data from a larger data set. Although oversampling consists of artificially increasing the size of a dataset by repeating some observations or creating new observations to reach the required size. In this article, we focus our work on the oversampling technique for the small group that represents the malicious user.

The technology used is KmeansSMOTE.

The KS-Catboost model contains two technologies, KmeansSmote and Catbooost. The first stage manages the imbalance of data to give the Catboost algorithm the ability to learn both harmful and normal classes well and eliminate the problem of overfitting the dominant class which represents the normal user. For this reason, the two algorithms were used together. The first one solves the data problem to provide the classifier with a great ability to learn and identify the harmful user. Catboost was also used for its superior ability to classify complex data.

A. KMeansSmote

The proposed model uses the popular and simple k-means clustering algorithm along with SMOTE oversampling to rebalance the data. By oversampling the safe regions, this method avoids the generation of noise. He is also interested in his work on the imbalance between classes and tackle the problem of small divisions by expanding minority areas Scattered also investigates imbalances within classes. This method was adopted because of its simplicity, ease, and widespread availability of Kmeans and Smote. This method is different from other methods, not because it is easy and simple, but rather because of its efficient way of distributing the artificial samples according to the density of each block.

The K-means SMOTE algorithm takes place in three stages: pooling, filtering and oversampling. The first step is to use Kmeans to group the input samples into K groups. Next comes the filtering step, in which you select the samples you want to multiply past the limit, and the samples from the minority class are selected. After that, artificial samples are generated whose distribution is sparse in each group, through the application of SMOTE, which generates within-group samples for minority groups that are less likely to generate noise, and the distribution of samples in the minority group is also balanced.

In the last oversampling step, SMOTE is used to adjust the proportion of minority and majority classes [20].

B. CatBoost

CatBoost [25] is a machine learning algorithm used for binary classification tasks. It is a variant of the AdaBoost algorithm designed specifically for unbalanced datasets, where one class is much more prevalent than the other.

CatBoost is a high-performance algorithm that can manipulate multiple data such as text, images, and audio to solve complex problems without extensive data training. Where the name of the algorithm is formed from the two words "Category" and "Boosting", it is an algorithm that uses decision trees to boost the gradient. This algorithm gives accurate results with relatively little data.

They are widely used to solve complex problems such as fraud detection.

The CatBoost algorithm works by iteratively training a sequence of weak classifiers, where each classifier is trained on a weighted version of the training data. The weights are adjusted so that subsequent classifiers focus more on the minority class, which helps improve classification performance on the unbalanced dataset.

During the training process, the algorithm assigns higher weights to misclassified samples from the minority class, which allows subsequent weak classifiers to pay more attention to these samples. In this way, the Catboost algorithm learns to focus on the minority class and ultimately creates a powerful classifier that can accurately classify minority and majority classes. Overall, Catboost is a useful tool for dealing with unbalanced datasets in binary classification tasks.

Here are the main steps of the CatBoost algorithm:

Data Preprocessing: CatBoost can handle categorical features directly, without the need for hot encoding or other preprocessing steps. It uses an algorithm called "Ordered Boosting" to transform categorical features into numerical values based on their target stats.

Tree Building: CatBoost builds a sequence of decision trees, where each tree attempts to correct errors from the previous tree. The trees are constructed using a process called "gradient-based one-sided sampling" which reduces overfitting and improves generalization.

Gradient Calculation: CatBoost uses a variation of gradient boosting called "Newton boosting" which uses second-order gradient information to improve the convergence speed and accuracy of the optimization process.

Regularization: CatBoost includes several regularization techniques such as "L1 regularization" and "random feature selection" to avoid overfitting and improve generalization.

Prediction: Once the model is trained, CatBoost uses the sequence of decision trees to make predictions on new data. It can handle both classification and regression problems and provides predictions in the form of class probabilities or continuous values.

Overall, CatBoost is known for its ability to handle categorical features, deliver high-accuracy results with minimal data preparation, and handle unbalanced datasets. In Fig. 3, the CatBoost algorithm [25] is explained.

Algorithm : CatBoost

input : $\{(\mathbf{x}_i, y_i)\}_{i=1}^n, I, \alpha, L, s, Mode$

1 $\sigma_r \leftarrow$ random permutation of $[1, n]$ for $r = 0..s$;
2 $M_0(i) \leftarrow 0$ for $i = 1..n$;
3 **if** $Mode = Plain$ **then**
4     $\quad\lfloor\ M_r(i) \leftarrow 0$ for $r = 1..s, i : \sigma_r(i) \leq 2^{j+1}$;
5 **if** $Mode = Ordered$ **then**
6     $\quad$ **for** $j \leftarrow 1$ **to** $\lceil \log_2 n \rceil$ **do**
7        $\quad\quad\lfloor\ M_{r,j}(i) \leftarrow 0$ for $r = 1..s, i = 1..2^{j+1}$;
8 **for** $t \leftarrow 1$ **to** I **do**
9     $\quad T_t, \{M_r\}_{r=1}^s \leftarrow BuildTree(\{M_r\}_{r=1}^s, \{(\mathbf{x}_i, y_i)\}_{i=1}^n, \alpha, L, \{\sigma_i\}_{i=1}^s, Mode)$;
10    $\quad leaf_0(i) \leftarrow GetLeaf(\mathbf{x}_i, T_t, \sigma_0)$ for $i = 1..n$;
11    $\quad grad_0 \leftarrow CalcGradient(L, M_0, y)$;
12    $\quad$ **foreach** $leaf\ j\ in\ T_t$ **do**
13       $\quad\quad\lfloor\ b_j^t \leftarrow -\text{avg}(grad_0(i)$ for $i : leaf_0(i) = j)$;
14    $\quad M_0(i) \leftarrow M_0(i) + \alpha b_{leaf_0(i)}^t$ for $i = 1..n$;
15 **return** $F(\mathbf{x}) = \sum_{t=1}^I \sum_j \alpha\, b_j^t \mathbb{1}_{\{GetLeaf(\mathbf{x}, T_t, ApplyMode) = j\}}$;

Fig. 3. CatBoost algorithm

4.5 Testing and Evaluation

This step evaluates the performance of the model and comes after the learning step. The data provided to the model is not the same as the data learned during the training step. The data is unbalanced, the harmful category is very little compared to the normal category, and the data submitted to the model in the testing phase has not been seen before, so that the assessment is accurate. It is noted whether the proposed model is capable of detecting the malicious user or not.

5 Results and Dissection

We used Python programming language with Tensorflow and keras as hidden interface to implement the proposed system. And to run the code, we used google colab to write and run any Python code.

5.1 Training/Test Split

To train the model, the data must be divided into a training and a test set. There are two different approaches, instance-based splits and user-based splits. In this article, we chose instance-based splitting, we took 80% of the data for model training and 20% for testing.

In Table 2 we show the number of instances for each class, where 0 represents the normal user and 1 represents the malicious user.

Table 2. Number of instances in each class

Class	0	1
Training set	263588	773
Testing set	589	193

The number of features extracted from the data is very large, so we organized the features according to their importance and focused our selection on the features that increase the accuracy of the model and we got the 80 features shown in Fig. 4.

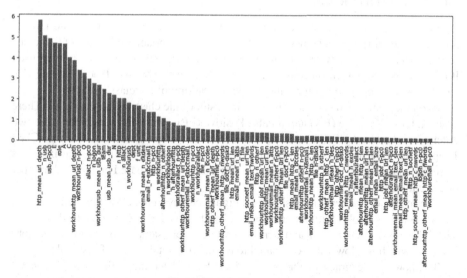

Fig. 4. Feature Importance

5.2 Tuning Hyperparameters

In our model we have fixed the following parameters to improve the results, classify it catboost, the number of iterations equal 6000 and learning_rate = 0.04, depth = 6. The KmeansSmote algorithm take a value of 1 for random_state and cluster_balanc_threshold = 0.0005, k_neighbors = 10, sampling_strategy = 1.

5.3   Evaluation Metric

As the data is highly unbalanced, we chose to use the commonly used measures of F-score and accuracy, recall and precision.

This is to see how accurate the model is in detecting the malicious user, which is a very small percentage compared to the normal user.

The mathematical representation of the rating scales used is shown below.

$$Accuracy = \frac{TN + TP}{TN + TP + FN + FP} \tag{1}$$

$$F-score = 2.\frac{Precision.Recall}{Precision + Recall} \tag{2}$$

$$Recall = \frac{TP}{TP + FN} \tag{3}$$

$$Precision = \frac{TP}{TP + FP} \tag{4}$$

5.4   Comparison of Results

Figure 5 represents the confusion matrix that contains the best result of the models trained on the same data. It shows the performance of the model used to detect malicious activity. This matrix shows results for both classes of malicious user and normal user. It relies in its assessment on four values, True Negative (TN) and True Positive (TP), which provide the highest values, indicating the validity of the classification.

If, on the contrary, the values are low, the result of the classification is unfair. There is also a False Negative (FN) and a False Positive (FP), promising values for these two criteria are decreasing values and the reverse is considered a misclassification. Class 1 represents TP and Class 0 represents TN. Each cell of the confusion matrix shows correctly and incorrectly classified samples for each category based on the total number of evaluation samples.

The results of the evaluation showed a high accuracy in the classification:

All normal samples without error were identified as TN = 65898 and FP = 0. For harmful activities TP = 174 and it misclassified a few samples of harmful category FN = 19 which is a low value compared to the number total of harmful samples.

To know the accuracy of the proposed model in detecting malicious activities, it should be compared to several other models run on the same cert v4.2 data and the resulting values should be compared. The algorithms used in the comparison are RandomForest, CatBoost, KmSmote with XGBoost and KmSmote with CatBoost. All these algorithms are applied to the same data and the same set of features extracted before. See Table 3.

We note that the highest accuracy is the accuracy of the proposed model, because it gave an accuracy = 99.97%, and since the data is unbalanced, we have to look at the value of the F1score, which is equal to 95%, which is the highest value compared to other models whose values range from 66% to 92%.

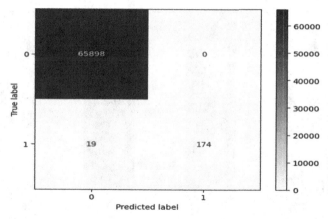

Fig. 5. Confusing matrix of KS-Catboost test

Table 3. Comparison KS-Catboost with other model

	Accuracy	F1-score	Precision	Recall	AUC
RandomForest	99.85%	66/20%	98.96%	49.74%	74.86%
CatBoost	99.95%	92.22%	99.40%	86.01%	93%
KmSmote + XGBoost	99.93%	88.31%	98.10%	80.31%	90.15%
KS-CatBoost	99.97%	95%	99.98%	90.15%	95%

The values of F1score and precision are close indicating the accuracy of the model in identifying benign and harmful activities with high efficiency.

Although KmSmote + XGBoost in this model addressed the data imbalance and gave a high accuracy of 99.93%, the F1 score result is 88.31% indicating its inability to accurately identify malicious activities. This also shows up in the recall result, which is 80.31%. This value shows a large number of malicious activities that have not been recognized and considered as a normal user. The highest value of Recall is equal to 90.15% recorded by the proposed model, indicating its good ability to identify the malicious user and the FN value is low compared to other algorithms.

RandomForest gave unsatisfactory results, the recall value is equal to 49.74%, which is the lowest value obtained, and the F1 score value is equal to 66.20%, indicating its inability to identify malicious activities. For CatBoost, it gave high values F1-score = 92.22% and Recall = 86.01% compared to other algorithms but it did not provide high accuracy like the proposed model.

For the other values, high recall scores equaled 90.15% and auc and accuracy gave their highest values, which concluded that the model provided high precision in detecting the malicious user. See Fig. 6.

In Table 4, we compare the proposed model with previous work, and relied on the selection of models using machine learning and deep learning algorithms. This work

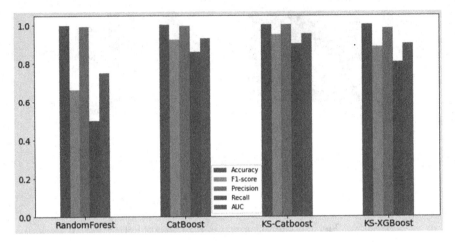

Fig. 6. KS-Catboost model detection accuracy.

Table 4. Comparison with previous work.

Model	Accuracy	F1-Score	Precision	Recall	AUC
[17]	99.94%	95%	–	99.66%	97.7%
[14]	99.92%	77.44%	64.60%	–	–
[9]	90.60%	94%	97.%	–	95.3%
[19]	99.47%	92.26%	99,47%	–	99.79%
[15]	96%	95%	–	–	95%
KS-Catboost	99.97%	95%	99.98%	90.15%	95.07%

addressed the data imbalance problem to find the malicious user due to the difficulty of finding samples with a small percentage.

There are also papers that used the boosting technique, which we relied on in our proposed model, and all to make an accurate comparison across multiple metrics. We have observed that the KS-Catboost model excels in detecting malicious users who represent the small percentage. However, the catboost algorithm was able to identify malicious activity, giving the highest accuracy of 99.97%. For F1-score, it gave a score of 95%, which is the highest recorded value, indicating that the model is able to predict both categories well and is not biased towards any category by relation to the other. By comparing all the results, we see the superiority of the model. He gave great precision in identifying the initiate and the new worker in the organization.

6   Conclusion

In this article, we aim to build an industrial system to detect insider attackers. We had issues extracting the appropriate functionality to represent user behavior within the organization. There is an imbalance in the data set because there are fewer malignant events than benign events. Therefore, we have proposed a KS-Catboost model that addresses this problem and uses the set learning algorithm for its simplicity and ability to classify despite the existence of unbalanced data.

The data used to test the model is CERT V 4.2. The final results compared to other modules and previous work yielded high accuracy in identifying the malicious user.

References

1. Gayathri, R.G., Atul, S., Xiang, Y.: Image-Based Feature Representation for Insider Threat Classification. Appl. Sci. **10**(14), 4945 (2020). https://doi.org/10.3390/app10144945
2. Figures: Insider threat statistics for 2022: facts and figures (2022). Ekransystem.com. Available: https://www.ekransystem.com/en/blog/insider-threat-statistics-factsand-figures. Accessed 05 Apr 2022
3. Verizon: 2019 Data Breach Investigations Report. In Computer Fraud & Security; Elsevier BV: Oxfordshire, UK, vol. 2019, p. 4 (2019)
4. Accenture/Ponemon Institute: The Cost of Cybercrime, Network Security; Elsevier BV: Amsterdam, The Netherlands, vol. 2019, p. 4 (2019)
5. IBM: Cost of a Data Breach Report 2019. In Computer Fraud & Security; Elsevier BV: Oxfordshire, UK, vol. 2019, p. 4 (2019)
6. Garcia, A., Orts-Escolano, S., Oprea, S., VillenaMartinez, V., Martinez-Gonzalez, P., Garcia-Rodriguez, J.: A survey on deep learning techniques for image and video semantic segmentation. Appl. Soft Comput. **70**, 41–65 (2018)
7. Yuan, F., Shang, Y., Liu, Y., Cao, Y., Tan, J.: Data augmentation for insider threat detection with GAN. In: 32nd International Conference on Tools with Artificial Intelligence, ICTAI 2000 (2020)
8. Azaria, A., Richardson, A., Kraus, S., Subrahmanian, V.S.: Behavioral analysis of insider threat: a survey and bootstrapped prediction in imbalanced data. IEEE Trans. Comput. Soc. Syst. **1**(2), 135–155 (2014)
9. Yuan, S., Wu, X.: Deep learning for insider threat detection: review, challenges and opportunities. Comput. Secur. **104**, 1–14 (2021)
10. Zhang, C., Wang, S., Zhan, D., Tingyue, Y., Wang, T., Yin, M.: Detecting insider threat from behavioral logs based on ensemble and self-supervised learning. Secur. Commun. Networks **2021**, 1–11 (2021). https://doi.org/10.1155/2021/4148441
11. AlSlaiman, M., Salman, M.I., Saleh, M.M., Wang, B.: Enhancing false negative and positive rates for efficient insider threat detection. Comput. Secur. **126**, 103066 (2023). https://doi.org/10.1016/j.cose.2022.103066
12. Yuan, S., Wu, X.: Deep learning for insider threat detection: review, challenges and opportunities. arXiv:2005.12433v1 (2020)
13. Raval, M.S., Gandhi, R., Chaudhary, S.: Insider threat detection: machine learning way. In: Conti, M., Somani, G., Poovendran, R. (eds.) Versatile Cybersecurity, pp. 19–53. Springer, Cham (2018). https://doi.org/10.1007/978-3-319-97643-3_2

14. Yuan, F., Cao, Y., Shang, Y., Liu, Y., Tan, J., Fang, B.: Insider threat detection with deep neural network. In: Shi, Y., Haohuan, F., Tian, Y., Krzhizhanovskaya, V.V., Lees, M.H., Dongarra, J., Sloot, P.M.A. (eds.) Computational Science – ICCS 2018: 18th International Conference, Wuxi, China, June 11–13, 2018, Proceedings, Part I, pp. 43–54. Springer, Cham (2018). https://doi.org/10.1007/978-3-319-93698-7_4

15. Al-Mhiqani, M.N., et al.: A review of insider threat detection: classification, machine learning techniques, datasets, open challenges, and recommendations. Appl. Sci. **10**(15), 5208 (2020). https://doi.org/10.3390/app10155208

16. Liu, L., de Vel, O., Chen, C., Zhang, J., Xiang, Y.: Anomaly-based insider threat detection using deep autoencoders. In: 2018 IEEE International Conference on Data Mining Workshops (ICDMW), pp. 39–48. IEEE (2018)

17. AL-Mhiquani, M.N., Ahmed, R., Abidin, Z.Z.: An integrated imbalanced learning and deep neural network model for insider threat detection. Int. J. Adv. Comput. Sci. Appl. **12**, 573–577 (2021)

18. Yuan, F., Shang, Y., Liu, Y., Cao, Y., Tan, J.: Data augmentation for insider threat detection with GAN. In: 32nd International Conference on Tools with Artificial Intelligence, ICTAI 2020 (2020)

19. Mohammed, M., Kadhem, A., Maisa, S., Ali, A.: Insider Attacker Detection Using Light Gradient Boosting Machine. Tech-Knowledge **1**, 48–66 (2021)

20. Douzas, G., Bacao, F., Last, F.: Oversampling for imbalanced learning based on K-Means and SMOTE. Inf. Sci. **465**, 120 (2017). https://doi.org/10.1016/j.ins.2018.06.056

21. Janjua, F., Masood, A., Abbas, H., Rashid, I., Zaki, M.M., Khan, M.: Textual analysis of traitor-based dataset through semi supervised machine learning. Future Gener. Comput. Syst. **125**, 652–660 (2021). https://doi.org/10.1016/j.future.2021.06.036

22. Glasser, J., Lindauer, B.: Bridging the gap: a pragmatic approach to generating insider threat data. In: Conference on Tools IEEE Security and Privacy Workshops (2013). https://ieeexplore.ieee.org/stamp/stamp.jsp?tp=&arnumber=6565236

23. Eldardiry, H., Bart, E., Liu, J., Hanley, J., Price, B., Brdiczka, O.: Multi-domain information fusion for insider threat detection. In: 2013 IEEE Security and Privacy Workshops, pp. 45–51 (2013). https://doi.org/10.1109/SPW.2013.14

24. Le, D.C., Zincir-Heywood, N., Heywood, M.I.: Analyzing data granularity levels for insider threat detection using machine learning. IEEE Trans. Netw. Serv. Manage. **17**(1), 30–44 (2020). https://doi.org/10.1109/TNSM.2020.2967721

25. Dorogush, A.V., Gulin, A., Gusev, G., Ostroumova Prokhorenkova, L., Vorobev, A.: Catboost: unbiased boosting with categorical features. arXiv preprint arXiv:1706.09516 (2017)

Author Index

A. Bennour et al. (Eds.): ISPR 2023, CCIS 1940, pp. 285–286, 2024.
https://doi.org/10.1007/978-3-031-46335-8

Printed in the United States
by Baker & Taylor Publisher Services